Markus Stoeppler (Ed.)

Springer

*Berlin
Heidelberg
New York
Barcelona
Budapest
Hong Kong
London
Milan
Paris
Santa Clara
Singapore
Tokyo*

Markus Stoeppler (Ed.)

Sampling and Sample Preparation

Practical Guide for Analytical Chemists

With 75 Figures and 46 Tables

 Springer

Dr. MARKUS STOEPPLER
Mariengartenstraße 1a
D-52428 Jülich
Germany

Updated and extended translation from the German.
Title of the German Edition:
Probennahme und Aufschluß (Springer Labormanual), published by:
Springer-Verlag Berlin Heidelberg New York, 1994

ISBN 3-540-61975-5 Springer-Verlag Berlin Heidelberg New York

Catalogin-in-Publication Data applied for

Sampling and sample preparation : practical guide for analytical chemists ; with 46 tables / Markus Stoeppler (ed.). – Berlin ; Heidelberg ; New York ; Barcelona ; Budapest ; Hong Kong ; London ; Milan ; Paris ; Santa Clara ; Singapore ; Tokyo : Springer, 1997
   Dt. Ausg. u.d.T.: Probennahme und Aufschluss
   ISBN 3-540-61975-5 (Berlin...)
   ISBN 0-387-61975-5 (New York...)
NE: Stoeppler, Markus [Hrsg.]

This work is subject to copyright. All rights are reserved, whether the whole or part of the material is concerned, specifically the rights of translation, reprinting, reuse of illustrations, recitation, broadcasting, reproduction on microfilm or in other ways, and storage in data banks. Duplication of this publication or parts thereof is permitted only under the provisions of the German Copyright Law of September 9, 1965, in its current version, and permission for use must always be obtained from Springer-Verlag. Violations are liable for prosecution act under German Copyright Law.

© Springer-Verlag Berlin Heidelberg 1997
Printed in Germany

The use of general descriptive names, registered names, trademarks, etc. in this publication does not imply, even in the absence of a specific statement, that such names are exempt from the relevant protective laws and regulations and therefore free for general use.

Coverdesign: de 'blik, Berlin
Typesetting: Fotosatz-Service Köhler OHG, Würzburg

SPIN: 10095586   52/3020-5 4 3 2 1 0 – Printed on acid-free paper

Preface

The significant progress achieved in modern instrumental analysis has led to a continuous lowering of detection limits and improved precision. This should in principle permit the reliable and extremely precise analysis of trace compounds mainly trace elements, at levels down to the lowest natural concentrations.

However, the frequently observed very high discrepancies between the analytical results of different laboratories as well as the deviations from true values are, regrettably, still common in analytical practice. Basic methodological errors at the determination step can usually be minimized or even avoided by carefully performed quality control measures – e.g. by interlaboratory comparisons and the proper use of certified reference materials. The most severe and often underestimated error sources, however, are those connected with the whole and often extremely complex sampling process, and also to a lesser extent, with sample preparation prior to analysis. Thus, for these initial steps of an analytical procedure particular experience is needed, as well as a detailed knowledge of the interrelations between these steps, which always have to be applied with the utmost care.

In collaboration with a number of very experienced colleagues working in different research areas, the editor of this book has tried to contribute to a better understanding of these particular error sources and how they can be overcome in a series of training courses held during the last decade at the "Haus der Technik", Essen, Germany. The condensed content of these courses, subdivided into the two main parts of sampling and sample preparation, was already presented in the German edition of this book, which appeared in 1994. Its remarkable success has led to the present updated and generally improved English edition in which some themes of primary importance have been somewhat enlarged for the sake of a broader international audience.

I do sincerely hope that this English version will be a versatile and valuable help for all those working in various scientific branches who in daily laboratory practice are frequently confronted with questions of how best to solve problems in sampling and sample preparation, in particular for trace and ultratrace elemental analysis.

Jülich, Autumn 1996 *Markus Stoeppler*

Contents

1 Sampling: an Introduction 1
MARKUS STOEPPLER

1.1 General Remarks 1
1.2 Error Sources Prior to Total Element Determination 1
1.3 Sampling for Subsequent Determination of Chemical Species 4
1.4 Studies to Evaluate Sampling Errors 5
1.5 References 6

2 Human Specimens 7
CORNELIA MÜLLER and ROLF ECKARD

2.1 Introduction 7
2.2 Human Specimens 7
2.3 Characterization 8
2.4 Sampling Procedures 9
    2.4.1 Whole Blood/Blood Plasma 9
    2.4.2 Urine 10
    2.4.3 Scalp Hair 10
    2.4.4 Human Milk 11
2.5 References 12

3 Wet Precipitation: Rain and Snow 13
PETER OSTAPCZUK

3.1 Introduction 13
3.2 Sampling Strategy 14
    3.2.1 Sampling Area 15
    3.2.2 Sampling Period 15
3.3 Sampling Equipment 16
3.4 Sampling in Winter 19
3.5 Sample Storage 20
3.6 Sample Preparation 21
3.7 Analytical Procedures 21
3.8 Results and Discussion 21
3.9 Conclusions 24
3.10 References 25

4 Sampling of Sea- and Fresh Water for the Analysis of Trace Elements   26
ECKARD HELMERS

4.1 Introduction 26
4.2 Sample Handling 27
    4.2.1 Preparatory Steps 27
    4.2.1.1 Clean Room Requirements and Behaviour of Personnel 27
    4.2.1.2 Selection of Labware and Sampling Bottles, Cleaning Procedures 28
    4.2.1.3 Purification of Water and Acids 29
    4.2.2 Contamination Effects: an Example 30
    4.2.3 Need for Filtration and Filtration Design 31
    4.2.4 Storage 34
4.3 Sampling Procedures 35
    4.3.1 Collection of Sea Water 35
    4.3.1.1 Water Column 35
    4.3.1.2 Surface Sea Water 35
    4.3.2 Collection of Fresh Water: Lakes, Rivers, Estuaries 36
4.4 Digestion 37
    4.4.1 Digestion of (Filtered) Water 37
    4.4.2 Digestion of Particulates 38
4.5 Typical Concentration Levels of Selected Trace Elements
in the Aquatic Environment and Suitable Analytical Methods 38
4.6 Quality Assurance During Analysis and Data Evaluation 40
    4.6.1 General Aspects 40
    4.6.2 Trend Monitoring: Decrease/Increase Verification 40
4.7 References 41

5 Soils and Soil Solutions 43
PIERRE DEL CASTILHO and RAINER BREDER (†)

5.1 Introduction 43
5.2 Materials 46
    5.2.1 Soil 46
    5.2.2 Soil Solution 47
5.3 Aspects of Soil Sampling and Recommendations for Its Realization ... 48
    5.3.1 Soil Sampling Report 51
    5.3.2 Details of Soil Solution Sampling 52
5.4 Sample Storage 53
    5.4.1 Soils 53
    5.4.2 Soil Solutions 54
5.5 Quality Control 54
5.6 Safety Precautions 55
5.7 References 55

6 Waste 57
ULRICH OSBERGHAUS and ECKARD HELMERS

6.1 Introduction 57
6.2 Theoretical Considerations for Sampling 58

		6.2.1	General Terms	58
		6.2.2	Deduction of a Criterion for Representativeness	58
		6.2.3	Relation Between Sampling Error and Analytical Error	60
		6.2.4	Variables which Affect the Sampling Error	60
		6.2.5	Estimation of the Required Number of Samples	62
		6.2.6	Examples	63
	6.3		Sampling and Storage	65
	6.4		Decomposition and Analysis	66
	6.5		Legal Requirements, Standards and Instruction Leaflets	67
	6.6		Specific Problems	69
		6.6.1	Monitoring Metal Concentrations in Municipal Waste and Incineration Residues	69
		6.6.2	Elemental Analysis of Sewage Sludge and Sewage Sludge Ash	71
		6.6.3	Metal and Metalloid Species in Gases from Sewage Sludges and Domestic Waste Deposits	72
	6.7		References	72

7 Collection, Preparation and Long-Term Storage of Marine Samples   74
Johann-Diederich Schladot and Friedrich Backhaus

7.1		Introduction	74
7.2		Tasks of the Environmental Specimen Bank	75
	7.2.1	Institutions Involved	75
	7.2.2	Selected Representative Ecosystems in Germany	76
7.3		Collection	78
	7.3.1	Collection Principles	78
	7.3.2	Collection of Marine Matrices	80
7.4		Sample Preparation	82
	7.4.1	Preparation of Homogenized Sub-Samples	82
	7.4.2	Cryogenic Grinding	83
	7.4.2.1	Sediments	83
	7.4.2.2	Bladderwrack	84
	7.4.2.3	Common Mussel	84
	7.4.2.4	Eelpout	84
	7.4.2.5	Herring Gull (Egg)	84
	7.4.3	Bottling of Homogenized Standard Sub-Samples	85
7.5		Conclusions	85
7.6		References	86

8 Biological Samples   88
Gerhard Wagner

8.1		Introduction: Ecological Basis – Information Content, Function and Indicative Value of Biological Specimens	88
8.2		Quality Assurance in Biological Sampling	89
	8.2.1	Repesentativeness of Biological Specimens and Samples	89
	8.2.2	Potential Errors in Biological Sampling	90

8.3	Development of Specified Sampling Plans		93
	8.3.1	Standardization of Sampling Procedures	93
	8.3.2	Definition and Delimitation of the Sampling Area(s)	94
	8.3.3	Necessary Sample Size and Confidence Intervals	95
	8.3.4	Sampling Time and Age of the Material	98
8.4	Specific Problems of Sampling and Preparation of Biological Environmental Specimens		100
	8.4.1	Difficulties of Sampling Caused by Abiotic and Biotic Factors	100
	8.4.2	Semi-Experimental Solutions of Sampling Problems	100
	8.4.3	Contamination or Deterioration of Plant Samples: the Problem of Washing	102
	8.4.4	Principles Governing the Sampling and Preparation of Animal Tissues for Environmental Analysis	104
8.5	References		105

9 Sampling of Industrial Material (Sampling for the Balancing of Elements in the Cement Industry) ... 108
WOLFRAM RECHENBERG and GEORG BACHMANN

9.1	Introduction		108
9.2	The Cement Clinker Burning Process		108
	9.2.1	Generalities	108
	9.2.2	Balancing	109
	9.2.3	Preheater Systems	110
	9.2.4	Sampling Points	111
	9.2.5	Determining Masses	112
	9.2.6	Frequency of Sampling	112
9.3	Sampling		113
	9.3.1	Conveyor Belts	113
	9.3.2	Pneumatic Conveyors	113
	9.3.3	Used Tires	115
	9.3.4	Fuel Oil	115
	9.3.5	Crude-Gas Dust	115
	9.3.6	Precipitated Dust	116
	9.3.7	Waste-Gas Dust	116
	9.3.8	Volatile Elements	117
9.4	Sample Preparation, Decomposition and Analysis		118
9.5	Checking Mass Flows		118
9.6	Conclusions		119
9.7	References		120

10 Food Products of Animal and Plant Origin ... 122
LOTHAR MATTER and MARKUS STOEPPLER

10.1	Introduction	122
10.2	Recommendations, Standards and Estimations of the Tolerable Intake of Toxic Metals via Food	123

10.3	Sample Collection	126
10.4	Sample Preparation	126
10.5	Homogenization	126
10.6	Decomposition Procedures	127
10.7	Analysis and Quality Control	129
10.8	Consumption Recommendations	130
10.9	References	131

11 Sample Preparation: an Introduction ... 132
MARKUS STOEPPLER

11.1	General Remarks	132
11.2	Error Sources	133
11.3	Decomposition Directly Prior to Determination in Analytical Systems	136
	11.3.1 Liquid Samples	136
	11.3.2 Solid Samples	136
11.4	Quality Assurance	139
11.5	References	139

12 Pressure Digestion: Apparatus, Problems and Applications ... 142
EWALD JACKWERTH, MICHAEL WÜRFELS
Translated by PHILIP H. E. GARDINER

12.1	Introduction	142
12.2	Liner	144
12.3	Body	145
12.4	Safety Devices	146
12.5	Heating System	147
12.6	Conditions for Digestion	147
12.7	Completeness of Sample Decomposition	149
12.8	References	152

13 Microwave-Assisted Decomposition ... 155
JUTTA BEGEROW and LOTHAR DUNEMANN

13.1	Introduction	155
	13.1.1 Fundamentals of Microwave Systems	155
	13.1.2 Fields of Application	156
13.2	Microwave-Assisted Decomposition Apparatus	157
	13.2.1 Safety Precautions	157
	13.2.2 Microwave-Assisted Digestion Techniques	157
	13.2.2.1 Low-Pressure Systems with Home Appliance Microwave Ovens	158
	13.2.2.2 Commercial Low-Pressure Microwave Systems	159
	13.2.2.3 High-Pressure Microwave Systems	159
	13.2.2.4 Non-Pressurized Microwave Systems	160
	13.2.2.5 Dry Ashing in a Microwave Oven	161

13.3 Comparison of Different Microwave-Assisted Digestion Systems 161
  13.3.1 Microwave-Assisted Digestion Systems 161
  13.3.2 Conventional Digestion Procedures 162
13.4 Criteria for the Evaluation of Microwave Digestion Systems 162
13.5 Conclusions 165
13.6 References 166

14 Decomposition Methods for the Electrochemical Determination of Elements 167
Peter Ostapczuk

14.1 Introduction 167
14.2 Determination without Sample Decomposition 169
14.3 Oxidative UV-Digestion 169
14.4 Mineralization by Oxygen 170
14.5 Open Wet Digestion 173
14.6 Pressurized Digestion 174
14.7 Conclusions 180
14.8 References 180

15 Decomposition of Materials for Trace Analysis in the Cement Industry 183
W. Rechenberg and G. Bachmann

15.1 Introduction 183
15.2 Decomposition Methods 183
15.3 Examples 185
  15.3.1 Leaching with Water 185
  15.3.2 Leaching with Acids 188
  15.3.3 Dissolution under Pressure 189
  15.3.4 Combustion 190
  15.3.5 Fusion 193
  15.3.6 Evaporation 194
15.4 References 196

Subject Index 199

List of Contributers

Dr. Georg Bachmann
Alsenbreitenburg Zement und Kalkwerk GmbH, P.O. Box 1160
D-25564 Lägerdorf, Germany

Friedrich Backhaus
Institut für Angewandte Physikalische Chemie (IPC),
Forschungszentrum Jülich, D-52425 Jülich, Germany

Dr. Jutta Begerow
Medizinisches Institut für Umwelthygiene an der Heinrich-Heine-Universität
Düsseldorf, Postfach 103751, D-40028 Düsseldorf, Germany

Dr. Rainer Breder (†)
Institut für Angewandte Physikalische Chemie (IPC),
Forschungszentrum Jülich, D-52425 Jülich, Germany

Dr. Pierre Del Castilho
DLO Research Institute for Agrobiology and Soil Fertility,
P.O. Box 129, NL-9750 Ac Haren, The Netherlands

Prof. Dr. Lothar Dunemann
Medizinisches Institut für Umwelthygiene an der Heinrich-Heine-Universität
Düsseldorf, Postfach 103751, D-40028 Düsseldorf, Germany

Dr. Rolf Eckard
Westfälische Wilhelms-Universität, Umweltprobenbank für Human-Organproben,
Domagkstraße 11, D-48129 Münster, Germany

Dr. Eckard Helmers
Chemisches Institut, im Amt für Umweltschutz, Staffelnbergstraße 81
D-70184 Stuttgart, Germany
formerly: Alfred Wegener Institute of Polar and Marine Research
Chemistry Section, Bremerhaven, Germany

Prof. Dr. Ewald Jackwerth
Auf der Bokkenbrede 39, D-44287 Dortmund, Germany

LM.-Chem. DI Lothar Matter
Chemisches- und Lebensmitteluntersuchungsamt, Wörthstraße 120,
D-47053 Duisburg, Germany

Dr. Cornelia Müller
Westfälische Wilhelms-Universität, Umweltprobenbank für Human-Organproben,
Domagkstraße 11, D-48129 Münster, Germany

Dr. Ulrich Osberghaus
BUC im Haus für Umwelt und Technik (HUT), Kirberichsdorfer Weg 6,
D-52066 Aachen, Germany

Dr. Peter Ostapczuk
Institut für Angewandte Physikalische Chemie (IPC),
Forschungszentrum Jülich, D-52425 Jülich, Germany

Dr. Wolfram Rechenberg
Forschungsinstitut der Zementindustrie, Tannenstraße 2,
D-40476 Düsseldorf, Germany

Dr. Johann-Diederich Schladot
Institut für Angewandte Physikalische Chemie (IPC),
Forschungszentrum Jülich, D-52425 Jülich, Germany

Dr. Markus Stoeppler
Mariengartenstraße 1a, D-52428 Jülich, Germany

Dr. Gerhard Wagner
Zentrum für Umweltforschung, FR 6.6, Biogeographie, Universität des Saarlandes,
Postfach 1150, D-66041 Saarbrücken, Germany

Dr. Michael Würfels
DMT, Energie, Umwelt, Franz-Fischer-Weg 61, D-45307 Essen, Germany

Chapter 1
Sampling: an Introduction

MARKUS STOEPPLER

1.1
General Remarks

Sampling for subsequent trace analysis is doubtless by far the most crucial step in an analytical procedure. If not properly planned and practically performed by using appropriate sampling tools [1] with the utmost care and expertise [2-6], total – systematic as well as random – errors for sampling can range from a small percentage to several orders of magnitude.

Depending on the analytical task and the material to be collected, sources of possible errors stem from many, often totally different, sources. These sources are treated in some detail in Chapters 2 to 10 for a number of analytically relevant sample types. Thus, this introductory chapter intends to draw the reader's attention to the facts and special problems common to the whole field of the sampling of biological, environmental and technical materials, and to add some general references that deal in depth with the matter addressed in the first part of this book.

1.2
Error Sources Prior to Total Element Determination

There are some typical errors that often occur in the course of planning and performing sample collection prior to trace analysis.

The first step of any analytical task is the proper definition of what is intended as far as the final result is concerned, from the analytical and also often the legislative viewpoint. This means that appropriate experience with the methods available in the laboratory, with the object itself, and with the element concentration or situation which has to be characterized (single person, species, collective, place, area, pollution level, etc.) have to be considered and a reliable sampling plan developed at the start. Particular error sources in performing these various tasks are summarized and discussed below with reference to the main themes that will be treated in the later chapters of this book.

- *Contamination* of all used sampling tools and vessels, as well as *adsorption* of the analytes on the surface of the collection and storage tools, has to be strictly avoided or as far as possible minimized for the collection of liquid, sometimes also solid, samples with very low or only slightly elevated natural levels. These analytically difficult matrices include e.g. blood, urine, human milk (Chapter 2), wet precipitation (Chapter 3), sea water, fresh water (Chapter 4), soil solution (Chapter 5) and some basic food materials (Chapter 10). Examples for the de-

Table 1.1. Probable median endogenous trace-metal levels or ranges in body fluids of nonexposed persons (values in µg/l)

Element	Urine	Serum/plasma	Whole blood	Remarks
Al	< 5	< 2	≤ 2	
As	< 10	< 1	≤ 1	Inorganic arsenic
Cd	≤ 0.5	< 0.05	≤ 0.5	Nonsmokers
	≤ 1	< 0.1	≤ 2	Smokers
Co	< 0.5	< 0.2	< 0.2	
Cr	< 1	< 0.5	< 1	
Hg	≤ 2	< 2	≤ 2	
Pb	< 10	< 1	< 50–250	Females lower
Tl	≤ 0.5	< 5	< 5	

Note: Significantly higher levels occur in cases of occupational exposure with typical values in the order of mg/l.

Table 1.2. Average element levels in sea water (S) and fresh water (F) without significant pollution

Element	≤ 10 ng/l	≤ 100 ng/l	≤ µg/l	> 1 µg/l
Al			(S?)	(F)
As				(S, F)
Be	(S)	(F)		
Cd		(S)	(F)	
Co		(S?)	(F)	
Cr			(S)	(F)
Cu			(S)	(F)
Hg	(S) below 1	(F)		
Mn		(S)		(F)
Ni			(S)	(F)
Pb		(S)	(F)	
Se	(S)		(F)	
Tl	(S)	(F)		
Zn		(S?)		(F)

tection of contamination from stainless steel needles and tubes made from various plastic and other materials are given by Bro et al. [7] (nickel and chromium), and Morita et al. [8] (chromium, manganese, iron and molybdenum). The latter workers applied ICP-MS for investigating contamination influences. Typical trace element levels in human body fluids, sea and fresh water are given in Table 1.1 and Table 1.2 and have been taken from various, predominately recent, sources [2, 9, 10]. From these data it is obvious that the subsequently applied determination methods also have to be extremely sensitive and reliable [11, 12, 13]. For the proper dissection of human and animal tissues (e.g. marine and freshwater fish, see Chapter 7) with trace element levels in the order of a few µg/kg or even less, the use of tools made from quartz glass, purified by rinsing with ultrapure acids, has proven to be very useful (Fig. 1.1).

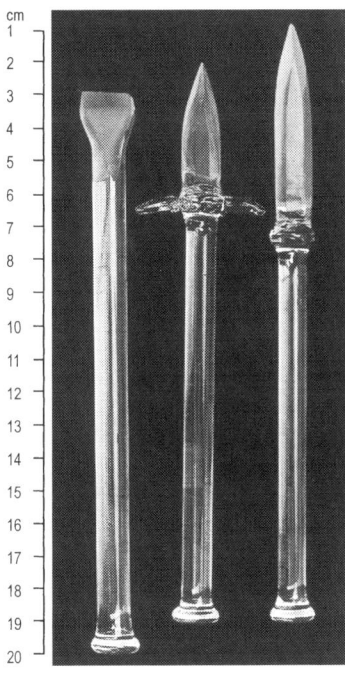

Fig. 1.1. Quartz tools for the contamination-minimized dissection of (fish) tissue

- Also the *time of sampling*, e.g. for body fluids [6, 14] and fresh waters (rivers, lakes) has to be considered. For waters it is also often useful to optimize the frequency of sampling to reduce the running costs of surveillance programmes [15].
- In many environmental samples, e.g. plants and soils, (Chapters 5, 7 and 8, see Table 1.3), waste (Chapter 6) and industrial materials (Chapter 9), the concentration of trace metals is often relatively high, e.g. at the mg/kg level. Thus, in these cases contamination plays no or only a very minor role. Here, the decisive terms for reliable environmental sampling are, for example, the selected (plant or animal) species considered as *bioindicators* [16] or *biomonitors* [17] and the homogenization procedures for reducing the basic material to manageable amounts in order to obtain representative analytical subsamples. This is especially important for heterogeneous materials that occur under various conditions (see Chapters 5, 6, 7, 9, 10 and 11). Furthermore, for environmental as well as for industrial materials the *place of sampling* is very closely linked to a proper sampling strategy (see for example the placing of wet precipitation samplers in Chapter 3, and suitable technical process positions for sampling industrial materials in Chapter 9).
- Last but not least, for all sampling tasks the careful selection of a *statistically relevant number and mass* of individual samples is very important and greatly depends on the distribution and concentration of the analyte in the collected material, i.e. its homogeneity or heterogeneity (see examples in Chapters 5, 6 and 8).

Table 1.3. Average element levels in plants (P) and soils (S), either without pollution or only slightly polluted

Element	≤ 100 µg/kg	≤ 1 mg/kg	≤ 10 mg/kg	> 10 mg/kg
Al				(P, S %)
As	(P)		(S)	
Be		(P)	(S)	
Cd	(P)	(S)		
Co			(P, S)	
Cr			(P)	(S)
Cu			(P, S)	
Hg	(P)	(S)		
Mn				(P, S %)
Ni			(P)	(S)
Pb			(P)	(S)
Se			(P, S)	
Tl	(P?)	(S)		
Zn				(P, S)

1.3
Sampling for Subsequent Determination of Chemical Species

A knowledge, or otherwise the best possible estimation, of the chemical form of the elements under investigation is of ever increasing importance for all analytical approaches in the fields of toxicological and environmental research and surveillance. However, collection procedures and techniques for many sample types are often not sophisticated enough for the preservation of chemical species and valency states until final determination. Thus, for the quantification of chemical species of elements special precautions are necessary in most cases. For liquid matrices the optimal approach is to perform any speciation analysis immediately after sampling [18]. Since this is often not feasible, pretreatment and storage procedures usually have to be applied. For aqueous matrices, the first step, preferably performed at the sampling site, is removal of particulate matter by filtration through 0.45 µm pore size membrane filters or by centrifugation so as to avoid changes in the distribution of chemical species due to various influences of the particulates. If storage of these solutions is necessary, they should be cooled down to approx. 4 °C in order to prevent additional rapid changes as a result of bacterial influences. It was shown that storage without acidification at 4 °C was effective for species preservation for some months. However, the appropriate preservation procedure depends on the chemical species to be determined [19, 20].

Speciation of metal-containing fractions of body fluids may be performed by various chromatographic methods, but also after denaturation of proteins prior to the chromatographic separation. An overview of sampling, sample preparation and sample preservation techniques for metal species is presented by Gardiner [21].

Solid biological materials should be prepared immediately after collection, for which separation of the parts is not required. For storage low temperatures are necessary. The separation of chemical species in tissues is usually performed after homogenization by solvent extraction, followed by specific separation/determination steps [22, 23].

Soils and sediments should preferably be stored at 4 °C after collection, but freezing might sometimes be also possible. For sediments, wet sieving and/or centrifugation in order to separate off interstitial water before storage and subsequent extraction/separation procedures is also common [24, 25]. A review of sample treatment and storage for organometallic compounds is given by Crompton [26].

1.4
Studies to Evaluate Sampling Errors

There is an urgent need for systematic studies as far as the error introduced during sampling procedures is concerned. Recently, the methodology of proficiency testing has been applied for the first time to sampling on contaminated land. Nine different sampler collected soil samples with the objective of estimating the mean concentration of Pb and Cu at a site. They used sampling protocols of their own choice, analysed their own samples and reported the results to the organizers of the study. Analytical bias was studied separately by distributing a matrix-matched reference material. The experiment demonstrated the feasibility of the proficiency test in sampling. Differences by a factor of 2.5 between the reported results illustrated that sampling can be a major source of error in the assessment of contaminated land [27].

Another approach in this field is the realization of a proposal for the "Comparative evaluation of European methods for sampling and sample preparation of soils" in the EU Framework Programme IV, Standards, Measurement and Testing. The project is approved by the European Commission and will be started soon. Similar to already established strategies and methods used in the analytical laboratory, the project will aid the development and use of suitable and effective tools for quality assurance (QA) and quality control (QC) in all the different phases of environmental studies.

The project is described as follows. "About fifteen participants, ideally one competent team from each EU-Member state, are requested to sample a test area according to their own national strategies, procedures and guidelines. The samples will be analysed centrally by an especially qualified laboratory. Detailed evaluation and comparison of results with comprehensive descriptions of the used sampling methods will indicate possible sources of systematic errors, restrictions in comparability and inappropriate, ineffective or defective strategies, methods or tools. An evaluation workshop will discuss these results and indicate solutions for improving the methodology. Two to three ensuing expert meetings will lead to pre-normative recommendations for national, European and international harmonisation of soil sampling procedures".

The author of Chapter 8 (G. Wagner) takes part in this study. He will certainly provide detailed information to all interested readers of this book.

1.5
References

1. Crosby NT, Patel I (1995) General Principles of Good Sampling Practice. The Royal Society of Chemistry, Cambridge, UK
2. Clarkson TW, Friberg L, Nordberg GF, Sager PR (eds) (1988) Biological Monitoring of Toxic Metals. Plenum, New York
3. Stoeppler M (1992) Sampling and Sample Storage. In: Stoeppler M (ed) Hazardous Metals in the Environment. Elsevier, Amsterdam, p 9
4. Hoffmann P (1992) Nachr Chem Tech Lab 40/12: M1–M32
5. Markert B (ed) (1994) Environmental Sampling for Trace Analysis. Verlag Chemie, Weinheim
6. Aitio A, Järvisalo J, Stoeppler M (1994) Sampling and Sample Storage In: Herber RFM, Stoeppler M (eds) Trace element analysis in biological specimens. Elsevier, Amsterdam, p 3
7. Bro S, Jorgensen PJ, Christensen JM, Horder M (1988) J Trace Elem Electrolytes Health Dis 2:31–35
8. Morita H, Kita T, Umeno M, Morita M, Yoshinaga J, Okamoto K (1994) Sci Total Environ 151:9–17
9. Merian E (ed) (1991) Metals and Their Compounds in the Environment. VCH Publisher, Weinheim
10. Herber RFM and Stoeppler M (eds) (1994) Trace element analysis in biological specimens. Elsevier, Amsterdam
11. Stoeppler M (1991) Analytical chemistry of metals and metal compounds. In: Merian E (ed) Metals and Their Compounds in the Environment. VCH, Weinheim, p 105
12. Howard AG and Statham PJ (1993) Inorganic Trace Analysis. John Wiley & Sons, Ltd., Chichester
13. Alfassi ZB (ed) (1994) Determination of Trace Elements. VCH Publisher, Weinheim
14. Versieck J (1985) CRC Crit Rev Clin Lab Sci 22:97–184
15. Johnson VM, Tuckfield RC, Ridley MN, Anderson AA (1995) Environ Sci Technol 30:355–358
16. Arndt U, Nobel W, Schweizer B (1987) Bioindikatoren, Möglichkeiten, Grenzen und neue Erkenntnisse. Eugen Ulmer Verlag, Stuttgart
17. Markert B (ed) (1993) Plants as Biomonitors. Indicators for Heavy Metals in the Terrestrial Environment. VCH Publisher, Weinheim
18. Völkening J and Heumann KG (1988) Fresenius Z Anal Chem 331:174–181
19. Batley GE (ed) (1990) Trace Element Speciation: Analytical Methods and Problems. CRC Press, Boca Raton FL
20. Quevauviller P, de la Calle-Guntinas MB, Maier EA, Camara C (1995) Mikrochim Acta 118:131–141
21. Gardiner PHE (1987) Topics in Current Chemistry, 141:145–174
22. Chau YK (1988) Sci Total Environ 71:57–58
23. Mazzucotelli A, Frache R, Viarengo A, Martino G (1988) Talanta 25:693–696
24. Kersten M, Förstner U (1990) In: Batley GE (ed) Trace Element Speciation: Analytical Methods and Problems. CRC Press, Boca Raton FL, p 245
25. Sager M (1992) Chemical speciation and environmental mobility of heavy metals in sediments and soils. In Stoeppler M (ed) Hazardous metals in the environment. Elsevier Science Publishers B.V., Amsterdam, p 134
26. Crompton TR (1988) Environ Int, 14:417–463
27. Argyraki A, Ramsey MH, Thompson M (1995) Analyst 120:2799–2803

Chapter 2
Human Specimens

CORNELIA MÜLLER and ROLF ECKARD

2.1
Introduction

Within the scope of the Federal Environmental Specimen Bank (see chapters 3, 7 and 8) specimens of human tissues are collected, characterized and stored at −85 °C in the Environmental Specimen Bank for Human Tissue, Münster (ESB Münster)[1]. The human material is obtained from autopsies (in the event of death by accidents) and healthy people.

The collection must inevitably take into account questions which have not even been thought of at present, as plying the simplest, reproducible and most practical techniques, which guarantee a minimum or lack of contamination. Even an excellent analytical procedure cannot compensate for bad sampling: "...unless the complete history of any sample is known with certainty, the analyst is well advised not to spend his time analyzing it" (Thiers, 1957). The two most important sources of errors are [2–4]:

- external contamination by sampling equipment (e.g. stainless steel cannulas, scissors);
- use of the wrong sampling technique (e.g. cannulas too small for blood plasma, separation of a cell fraction too late).

Element losses due to surface adsorption are only of secondary concern for the real sampling.

Consequently "Standard operating procedures" (SOPs) were established for use by the ESB Münster and are obligatory for sampling and the analytical procedures of all matrices.

2.2
Human Specimens

In this paper a selection of matrices from the Real-Time Monitoring (RTM) program is presented (sampling twice a year from a constant group of students). These specimens are selected because of their clinical and environmental/toxicological relevance: whole blood, blood plasma, urine, scalp hair and, as special "juice", human milk.

---

[1] This project is supported by the Federal Environmental Agency and the Federal Ministry of the Environment, Nature Protection and Nuclear Safety.

2.3
Characterization

As mentioned above, the samples were characterized by fixed procedures. These were: (a) a detailed anamnestic questionnaire; (b) chemical analyses: organic (for organochloropesticides, pentachlorophenol, polychlorinated biphenyls), inorganic (for

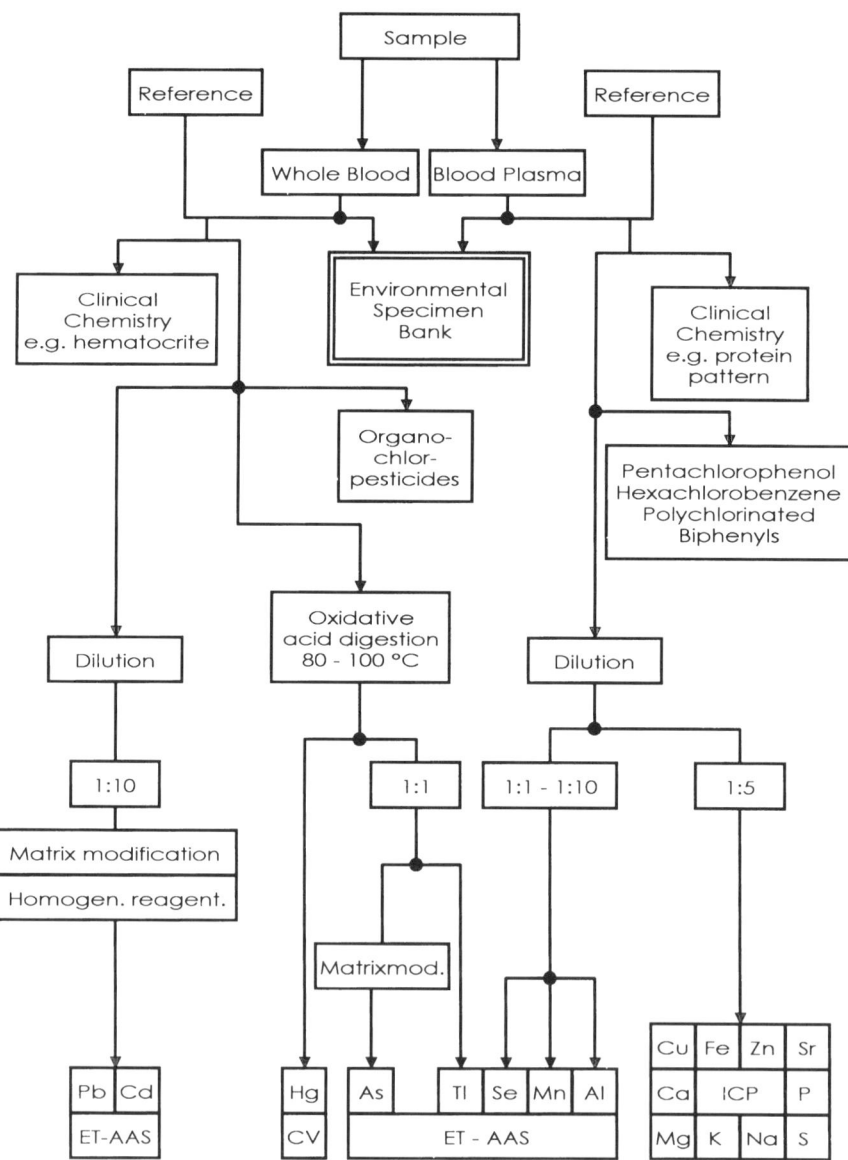

Fig. 2.1. Analytical scheme for whole blood and blood plasma

bulk and trace elements) and clinical chemistry (e. g. hematocrite, creatinine). Figure 2.1 shows the comprehensive analytical scheme for whole blood and blood plasma.

The analytical characterization may lead to a reliable database, allowing reference ranges to be established, e. g. for accidental and essential trace elements.

2.4
Sampling Procedures

The following descriptions are extractions from the above-mentioned SOPs, which have been proven in practice. Sampling tubes for inorganic characterization are always cleaned with 8% nitric acid and bidistilled water; for organic analysis petrolether of nanograde quality is used. The drying of sampling tubes is done under clean-room conditions (class 100).

For the collection of human milk both cleaning steps are first carried out for the sampling tubes and all materials which may come into contact with the milk.

Sampling vessels of different size (13.5, 30 and 50 ml) with screw caps and made of polypropylene were used. They are sufficiently stable at $-85\,°C$ and have no surface activity. Sampling equipment, which cannot be cleaned (e. g. sterilized syringes for single use), is checked for possible contamination at each new charge.

2.4.1
Whole Blood/Blood Plasma

Blood sampling is always done in the supine position, because sitting or standing may lead to a volume redistribution of the blood compartment [2]. Only non-powdered examination gloves should be used.

A tourniquet is applied lightly, the skin is cleaned with an alcohol swab (70% isopropanol). The venipuncture is done with needle of an infusion set that has a large diameter (19G). This will guarantee protection of the vessel and lower the risk of haemolysis.

All together ca. 80 ml blood are taken and subsampled immediately in heparinized vials (1 or 2 drops of heparine [5000 I. U./ml] per vial). The first 20 ml are drawn with a normal plastic syringe and reserved for organic analysis. For trace-element analysis in plasma, the blood is dripped directly into a prepared vial.

Scheme: 1. 20 ml drawn with 20 ml syringe (organic analysis)
2. 10 ml dripped (plasma trace elements)
3. 20 ml drawn with same syringe (inorganic analysis)
4. 10 ml dripped (plasma organic analysis)
5. 10 ml dripped/drawn (plasma organic analysis).

The tubes were closed directly after sampling and slightly shaken in order to distribute the anticoagulant evenly. Ethylendiamintetraaceticacid (EDTA), another frequently-used anticoagulant, presented a lot of problems during organic analysis. As EDTA is an excellent complexing agent for kations there is always risk of contamination. Regarding the above-mentioned scheme, falsification of a trace-element pattern in blood plasma as a result of contamination by the steel cannula (Cr, Ni, Co, Mn), can be avoided [4, 5]: the first subsample (for organic analysis) may be regarded as washing step for the following inorganic subsamples. To avoid haemolysis, it is recommended that the cell fraction is separated at the latest 2 hours after sampling (centrifugation for 10 min at 3000 rpm). This is important for iron and zinc as essential trace elements which have a higher intracellular concentration. Plasma is transferred (with pipette tips rinsed once with bidistilled water) to prepared tubes under clean-room conditions. The advantage of using blood plasma instead of blood serum is that the protein fraction is more reproducible (there has not yet been any precipitation of fibrinogen).

2.4.2
Urine

Urine is sampled according to detailed instructions ("man is possibly the most serious contamination source..." [6]) over 24 hours into pre-cleaned 2.5-l bottles. Samples should not be taken from women during menstruation, men should be warned that there is a risk of contamination by zinc in sperm remains. To guarantee the stability of the urine, the time that elapses between sampling and analysis should be kept as short as possible. To avoid the risk of contamination the urine is not acidified. After thorough shaking, the urine is divided up and the subsamples are frozen immediately. Checking density and creatinine content allows the collecting period to be controlled: collected or spontaneous urine?

2.4.3
Scalp Hair

In contrast to the sampling of other specimens, there has been some controversy about the sampling and pre-analytical handling (especially the washing procedure) of scalp hair.

For the ESB Münster the following procedure has been developed:
For cosmetic and physiological reasons – hair growth varies according to location on the head – a strand of hair (ca. 0.5 – 1 g) is cut off directly at the skin in the occipital region (Fig. 2.2). Afterwards the strand is divided into segments, each 2.5 cm

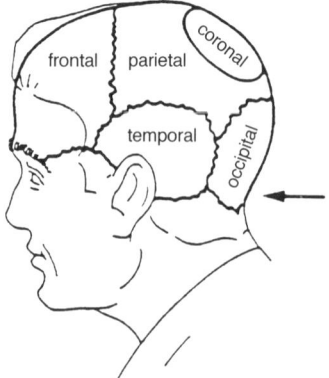

Fig. 2.2. Sampling sites for scalp hair (after [7])

long. Only the segment that was the nearest to the scalp is used for analysis and comparison, because (a) hair length varies considerably, and (b) the external contamination of the more distal segments cannot be differentiated. The longitudinal distribution of the trace elements may give information on the personal history of the individual being tested.

As the components of stainless steel (e.g. Cr and Ni) may be of interest in an element analysis normal scissors should not be used. Zirconium-ceramic scissors are the best alternative at present. They are sharper and less expensive than titanium scissors.

2.4.4
Human Milk

The sampling of human milk is always done with the assistance of a trained person. The breast, in particular the nipple, is cleaned with a sterile compress and bidistilled water, in order to remove possible contaminants (e.g. breast cosmetics). The sampling equipment, which includes an electric milk pump (Fig. 2.3) is unpacked and installed at the place where the sampling is done. The breast adaptor is pressed on to one breast, so that no air is drawn in and a light vacuum is created. The pump is started and controlled by the test person herself. As soon as the desired volume (ca. 30–50 ml) has been reached, the pump is stopped and the filled milk bottle is removed from the breast. To minimize contact with the air, the sample is rapidly subdivided into prepared tubes. If possible this procedure should be done under clean-room conditions in the laboratory.

If the ejaculation of the milk does not take place, even though the breast is full, the test person may use a nasal spray of oxytocine in order to stimulate the milk flow.

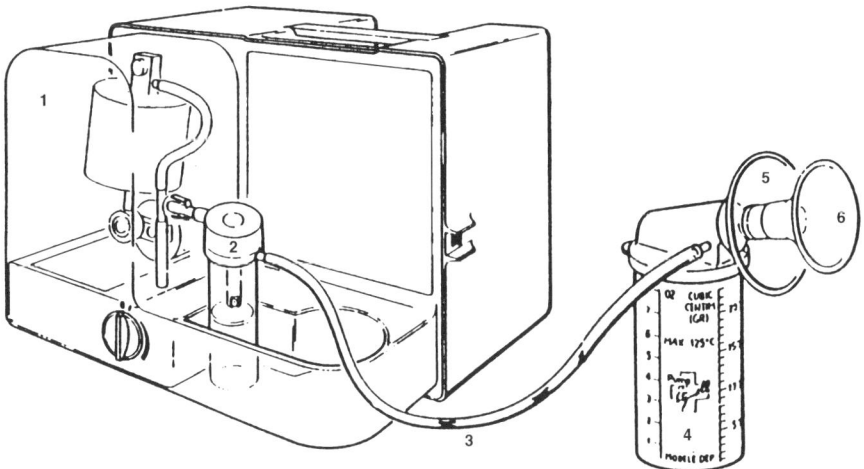

Fig. 2.3. Sampling equipment for human milk. 1: milk pump; 2: overflow vessel; 3: connection tube; 4: milk bottle; 5 and 6: breast adaptor

2.5
References

1. Thiers RE (1957) Contamination in Trace-Element Analysis and its Control. In: Glick E (ed) Methods of Biochemical Analysis. Interscience, New York, Vol 5
2. Behne D (1980) Problems of Sampling and Sample Preparation for Trace-Element Analysis in the Health Science. In: Brätter P, Schramel P (eds) Trace-Element Analytical Chemistry in Medicine and Biology. Walter de Gruyter, Berlin New York
3. Speecke A, Hoste J, Versieck J (1976) Sampling of Biological Material. In: LaFleur PD (ed) Accuracy in Trace Analysis: Sampling, Sample Handling, Analysis. U.S. Government Printing Office, Washington, Vol 1
4. Versieck J, Cornelis R (1989) Trace Elements in Human Plasma or Serum. CRC Press, Boca Raton
5. Brø S, Jørgensen PJ, Christensen JM, Hørder M (1988) J Trace Elem Health Dis 2:31
6. Stoeppler M (1991) Analytical Chemistry of Metals and Metal Compounds. In: Merian E (ed) Metals and Their Compounds in the Environment. Verlag Chemie, Weinheim New York Basel Cambridge, p 105
7. Pecoraro V, Astore JPL (1979) Messungen des Haarwachstums. In: Orfanos CE (ed) Haar und Haarkrankheiten. Gustav Fischer Verlag, Stuttgart New York

Chapter 3
Wet Precipitation: Rain and Snow

Peter Ostapczuk

3.1
Introduction

The geographical distribution of precipitation on the globe is determined by the general atmospheric circulation and by local topography, particularly the locations of mountains and oceans. In general, for condensation and precipitation of water, there must be saturated water vapor in the air and air must cool down. Over the clobe as a whole, three precipitation zones exist: one centered on the equator and the others centered around 50° N and 50° S [1]. On a smaller scale, mountains tend to cause a special effect: warm moist air carried onshore from the ocean rinses along a mountain front, causing precipitation on the windward side. The air is dry when it reaches the leeward side of the mountains and a dry inland area results. A monsoonal circulation, which occurs along the coast of continents, particularly in south-east Asia, results in extremely heavy but very seasonal precipitation.

Even when the air is supersaturated with water vapor, condensation to liquid water droplets will not occur unless small aerosol particles or cloud condensation nuclei are present. For rain to occur, these small cloud water droplets must grow, becoming larger and heavy enough to fall through the air without evaporation (Fig. 3.1). Rain is one of the most important agents for cleaning the atmosphere from anthropogenic and natural contaminants.

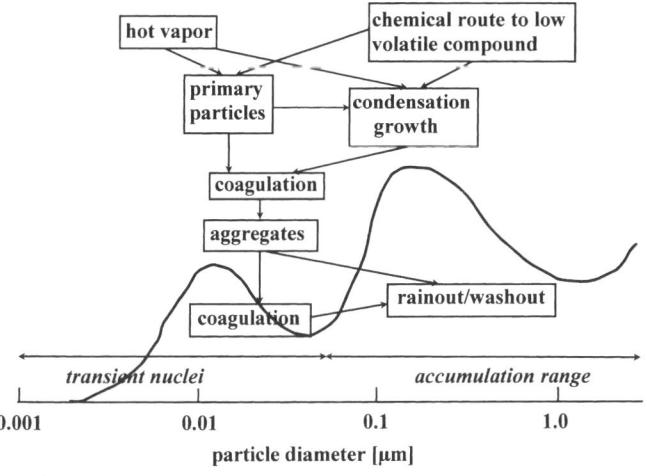

Fig. 3.1. Schematic diagram of the size distribution (expressed as surface area per increment in log of particle diameter) and formation mechanisms for atmospheric aerosols

In many cases it is necessary to define what rain is. There are no general criteria for this definition. Very small water droplets (diameter < 0.5 mm), without significant terminal, velocity, constitute what is known as fog [2]. Temperature changes in the atmosphere can evaporate the droplets without any precipitation or the droplets can condense to form drizzle or dew. This form of precipitation can be collected only by active samplers, if appropriate amounts of samples are to be collected. The rain intensity can vary from low, through moderate and strong to heavy.

3.2
Sampling Strategy

Water vapor is pure water. Since rain forms as a result of condensation on small aerosol nuclei, these become dissolved as ions in the rain. Sea salt and soil dust aerosols are the major natural contributors to dissolved components in rain [13]. In industrial areas rain is contaminated with heavy metals. Atmospheric gases dissolve in rain and react to form new species. When distilled water is in equilibrium with atmospheric CO_2, hydrogen ions – formed by the dissociation of carbonic acid – create a natural acidic rain with pH = 5.7. 80% of the atmospheric SO_2, which reacts in the atmosphere to produce fine aerosol particles of sulphuric acid that dissolve in the rain, are from anthropogenic sources (power and heating). Nitrate in rain comes from NO_x (75% anthropogenic) that is produced by fossil fuel burning and automobile exhaust. SO_2 and NO_x are the main contributors to acid rain. The sources of the ammonium ion in rain are largely agricultural, and nearly 90% of it is estimated to be anthropogenic (Table 3.1).

The collection of rain samples is essential for various reasons. It is necessary to know the background concentration of pollutants in the atmosphere for proper legislation. The time and space distribution of the pollutants can be very impor-

Table 3.1. Annual average concentration (µeq/liter) of chemicals in precipitation

Chemical constituent	Nova Scotia [20]	New Hampshire [2]	Antwerp [21]	Darmstadt [18]	Groß-Rohrheim [18]
H^+	29.9	69.3 ± 2.1	95 ± 68	37	16
NH_4^+	4.2	10.6 ± 0.6	157 ± 158	55.2 ± 3.5	33.3 ± 3.3
Ca^{2+}	4.3	6.5 ± 0.8	–	1.8 ± 0.2	1.0 ± 0.1
Na^+	26.1	4.8 ± 0.5	–	16.5 ± 1.2	13.0 ± 0.7
Mg^{2+}	2.9	3.0 ± 0.5	–	0.8 ± 0.1	0.5 ± 0.1
Al^{3+}	< 0.1	1.8 ± 0.2	–	–	–
K^+	1.1	1.5 ± 0.3	–	0.9 ± 0.1	0.8 ± 0.1
SO_4^{2-}	27.5	54.0 ± 2.1	209 ± 132	15.0 ± 0.2	9.1 ± 0.1
NO_3^-	9.7	23.5 ± 1.0	81 ± 83	23.4 ± 1.2	14.5 ± 0.5
Cl^-	29.5	11.2 ± 1.2	303 ± 403	12.3 ± 0.9	9.1 ± 0.8
PO_4^{3-}	–	0.3 ± 0.1	–	–	–
HCO_3^-	–	0.1	–	–	–
pH	4.6	4.16	4.2 ± 0.5	4.43	4.81

tant for governmental activities. From the scientific point of view, questions concerning chemical reactions in the atmosphere, modeling and the influence to the biosphere can be important [15]. For each of these questions, the development of a sampling strategy and full analytical procedures are the most important steps, before starting any experimental exercise.

3.2.1
Sampling Area

Depending on how the problem is defined, the sampling area is chosen. There are various restrictions in positioning the sampler. These restrictions are very strongly correlated to the problem definition. In many cases a compromise between the "possible" and the "required" must be done under field conditions. The following parameters can influence precipitation sampling: wind direction and speed, air temperature and moisture, topographical and geological conditions on sampling area. It can be estimated that in difficult topographical and geological situations, the rain collectors are not representative of a large sampling area. Collected rain represents only an area close to the sampler.

Some of the parameters mentioned above are not under control. In a real situation, the need for electrical connections, the use of electronic devices for parameter control, and a change of sampling containers impose limitations on the choice of an optimal sampling area. In many cases the sampling area is close to some natural or man-made hindrances. The distance from the sampler to these hindrances must be as large as possible. If heavy metals are to be determined in rainwater, then contamination by dust is a problem. To reduce this kind of contamination it is necessary to place the sampler at a minimum altitude of 150 cm [3]. This is a compromise between the convenience of changing the containers of sampled water and protection from contamination by dust. In windy conditions, an altitude of only 150 cm will not protect the collected sample from dust contamination, and higher elemental concentrations have been observed, especially after short rainfall.

3.2.2
Sampling Period

Depending on the capability of the analytical method and the nature of the problem, appropriate sampling periods can be chosen. There are samplers on the market, which can collect the rain samples within a period of several minutes. In such a cases the volume of the rain sample is limited and very much depends on the rain density. Furthermore, due to unfavorable relationships between the volume of the sample and the surface area of the collecting containers there is a high risk of contamination or loss of rain constituents. Errors can also be introduced if the water sticks to the inner wall of the funnel, thus not entering the sampling container. This so-called wetting loss depends on the total surface area of the funnel relative to its opening cross-section (Fig. 3.2). On the other hand, the short time period for sampling protects the rain samples from chemical or biological changes.

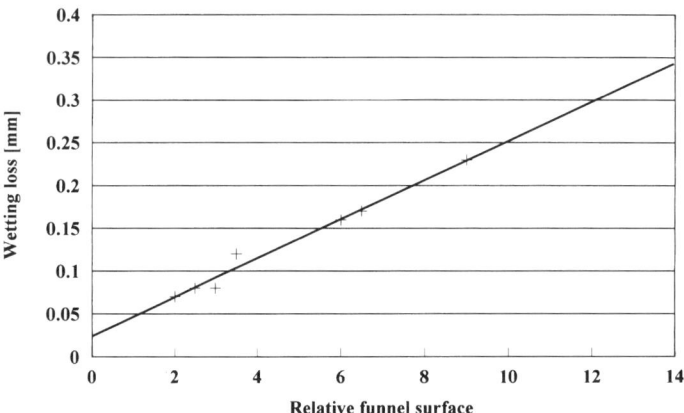

Fig. 3.2. Wetting loss in the funnel as a function of relative funnel surface (relationship between the whole surface and the total wall surface)

A sampling period of a few hours or a full day is very often chosen. In this case, it is necessary to change the sample containers after an appropriate period of time in order to guarantee the comparability of all collected samples. Because large numbers of rain samples are collected, the short sampling times are not convenient for long-term and network monitoring. For this kind of study a sampling period of one week has been estimated to be the best [4].

3.3
Sampling Equipment

The sampling equipment must be chosen according to to the definition of the problem. There are many samplers for monitoring the inorganic constituents of rain [16]. An intercomparison of 20 wet-only samplers [5, 6] have demonstrated that the precipitation sensors responsible for the opening and closing of the collectors have a marked influence on the composition of the samples. High sensitivity collectors also collect very small droplets like smog containing a lot of dust particles. Collectors with poor sensitivity do not collect the first raindrops, which are often highly concentrated. For this reason, the ion concentrations found in samples collected by these instruments were on average significantly lower than those for instruments with sensors of high sensitivity. One of the possible designs of rain samplers is shown in Fig. 3.3.

This wet-only sampler was developed at the Institute of Applied Physical Chemistry, at the Research Center in Juelich and produced in small series. The sensor is heated and the reaction time required to open or close the sampler can be regulated between 0.3 to 3 min. Due to the incorporated heating system it is also possible to collect snow samples. The water collected is immediately filter through a 0.45 µm filter. All parts of the sampler are made of polyethylene or polyurethane.

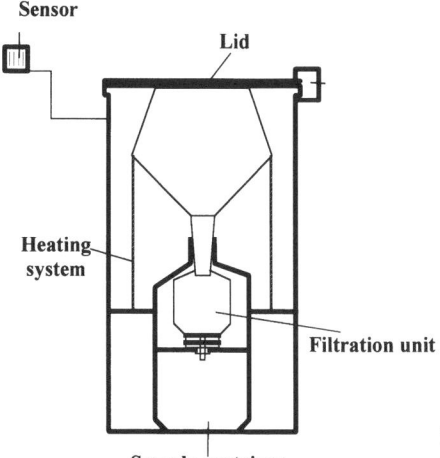

Fig. 3.3. A wet-only rain sampler with filtration unit

Rain samplers for monitoring organic or both organic and inorganic pollutants have also been developed [5]. In each of these samplers the following points are of importance:

- the amount of samples needed for all analytical steps;
- the accuracy of the volume of the sampled rainwater;
- contamination free sampling;
- the necessity or possibility of filtering the rain;
- the stability of rain sample in the container used with time.

For solving problems such as the investigation of trends, the extent to which the collected precipitation samples are representative of the sampling area and the accuracy with which the concentration of pollutants can be determined are very important. Without the correct knowledge of these parameters, it is difficult to calculate the global deposition rates of pollutants.

The volume of the rainwater sample collected is another important parameter for calculating concentration of pollutants in rainwater and the total deposition into the terrestrial ecosystem. The volume of rainwater can be estimated in two ways. The volume of water collected in the rain sampler is measured after each sampling period. To control this volume, a Hellmann sampler (used for determination of rainwater volumes by the German Weather Office) is placed close to the rain sampler. Figure 3.4 presents a comparison between the rainwater volume found by the KFA rain sampler and the Hellmann sampler. In some special cases a difference of more than 10% can be observed. For long-period observations good correlation between both sets of values was found.

Significantly large differences in the collected rainwater volume were observed when two samples were situated in the same sampling area but separated by a distance of few kilometers (Fig. 3.5a). In some special cases, the difference between the collected rain volumes was more than 100%. These differences are strongly

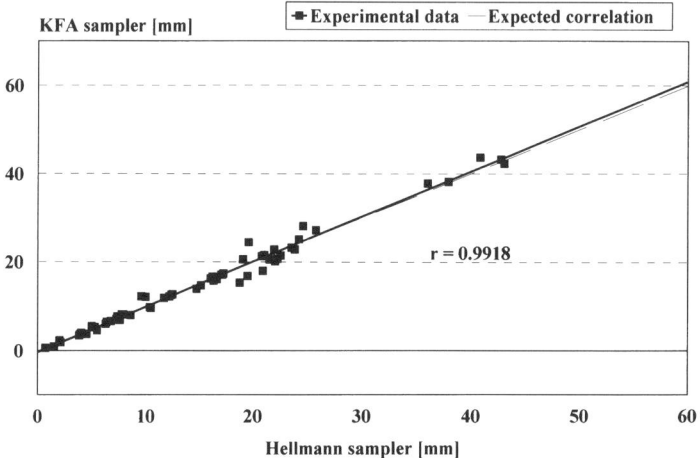

Fig. 3.4. Correlation between the volume of rain samples collected by a wet-only rain sampler and a Hellmann sampler

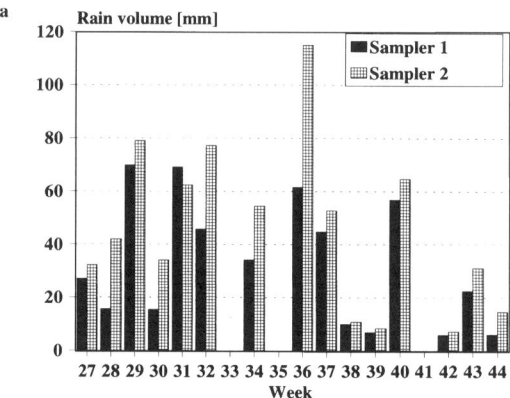

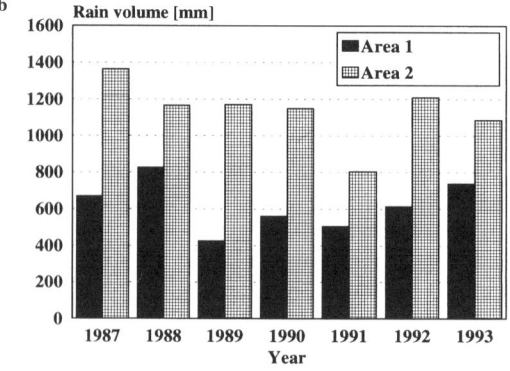

Fig. 3.5 a, b. Comparison of rain volumes collected by two wet-only samplers kept at a distance of few kilometers in two sampling areas: **a** weekly samples, **b** annual means

dependent on the topographic conditions in the sampling areas. Long-term observations have demonstrated that the total annual rainwater volume also changes significantly. The difference between the maximum and minimum values is more than 50%. A similar deviation in the annual rainwater volume is observed between two sampling areas (Fig. 3.5b).

The equipment used can sometimes contaminate the samples because the precipitation comes into contact with unsuitable materials. Contamination problems during rain sampling can be minimized by using materials which are not analyzed as pollutants. All such contamination problems can be easily avoided if the origin of the contamination is localized. New equipment needs time for cleaning under field conditions. Depending on the analyte and on the materials used, the time needed for cleaning can be several weeks.

Filtration is often recommended before monitoring the heavy metals in rain samples [6–9]. If the sample is not filtered, the particulate matter forms a sediment in the container and it is very difficult to transfer quantitatively all of the sample into the analytical equipment. The most important argument for immediate filtration is the long-term interaction between insoluble particles and the dissolved trace compounds.

Stability of the pollutants in the rain sampler can significantly influence the observed concentrations [10]. Experiments need to be performed systematically in order to determine the amount of pollutant adsorbed on the walls of the sampler and the sample container. Biological activity and chemical reactions in the rain sample are the other two of the many possibilities which can change the pollutant concentration in the sample.

3.4
Sampling in Winter

Snow collection with rain samplers is very difficult. For improving the collection of snow, and to prevent collected snow from being blown out again, deep funnels with a large surface area are used. With these sampler, it is not possible to filter the sample. If the weather changes and the snow begins thaw, the problems with unfiltered samples appear. Such instruments may have a high wetting loss. The chemical reactions which are possible during the thawing of snow can also change the equilibrium between the solid and liquid phases because of the increased concentration of hydroxy ions at the beginning of thawing. It is possible to heat all the samplers to enable the sample to be filtered. In many cases the sample volumes obtained by these samplers are significantly lower than those obtained by Hellmann samplers. This is due to the evaporation of water from the funnel and from the sample container.

The simplest way to collect snow is "sampling by hand". Similar to the collection of water samples, it is possible (in areas free of human activity) to collect the surface snow using appropriate containers. When using this kind of sampling, the main objective is to prevent contamination of the sample by human beings and the equipment used. The collected sample represents only a short period of time and a small sampling area.

The investigation of the occurrence of heavy metals such as Pb, Cd, Hg, Cu and Zn in the successive by dated snow and ice layers deposited in Antarctica and Greenland provides a unique opportunity to reconstruct the past and present-day changes in the large-scale tropospheric fluctuations of these toxic metals. Data for ancient ice, whose age can extend back to a few hundred thousand years, will indeed allow us to determine the natural, pre-human fluctuations of the past, hence establishing a firm reference against which to evaluate modern trends. Once these references are established, data for snow or ice deposited during the last centuries, especially since the industrial revolution, will allow us to estimate directly the human impact on these fluctuations both in the northern (Greenland data) and southern hemispheres (Antarctic data).

Such studies have unfortunately been hampered because the concentrations of heavy metals found in polar snow and ice are the lowest detected on earth. As an illustration, typical concentrations of Pb and Cd in Antarctic ice several thousand years old are as low as about 0.4 pg Pb/g and 0.1 pg Cd/g, respectively [12], i.e. concentrations which are much lower than those in the ultrapure water produced in most research laboratories. Consequently, many of the published data are erroneous by up to several orders of magnitude because of inadequate control of contamination during field sampling or/and laboratory analysis. It is only during the past few years that procedures have been developed that are refined enough to obtain reliable, albeit very incomplete, data [19].

In the near future, the most urgent need will be to obtain reliable and detailed time series to cover the past few hundred years both in Antarctica and in Greenland at single sites remote from local contamination sources. The involved scientists however, have to be able to demonstrate that the outset of their data are fully reliable: it will be mandatory to study the variation of the concentrations of the metals investigated from the outside to the inside of each individual sample, and to perform detailed and accurate blank determinations for the entire analytical procedure from sample decontamination to final analysis.

3.5
Sample Storage

To protect the rain samples from chemical changes before determination, it is necessary to store them. If the trace elements are of interest, the sample must be acidified to pH 1. Sample acidification can be done only after the sample filtration. If an unfiltered sample is acidified, the concentrations of trace elements found in the liquid phase is significantly higher than those observed in the same sample when acidified after filtration. Acid mobilizes the elements bound to particles and due to significant concentration differences between the dust particles and the rain sample, the latter may be "contaminated". After acidification, the samples are more or less stable and no significant changes in the element concentrations over 8 months were observed. For long-term storage it is be better to store the samples at $-30\,°C$.

If anions are also to be determined, then it is useful to divide the rain sample into two subsamples: one acidified for the determination of trace elements and the other, without any chemical pretreatment, for the determination of inorganic anions. If only SO_4^{-2}, NO_3^- and Cl^- are to be determined, the sample can be stored for several weeks at 4 °C.

3.6
Sample Preparation

For the determination of trace elements by anodic stripping voltammetry, it is necessary to digest the rainwater samples. For water samples, UV irradiation is the best digestion method (see Chapter 14 "Decomposition methods..." in this book).

If the rainwater sample is filtered, it is also necessary to digest filter with suspended matter. Digestion in an oxygen plasma is used when a lot of samples are to be digested at the same time. Total digestion time is 8 hours and can be done overnight. For individual samples, microwave or high pressure (HPA) digestion with nitric acid can be used.

3.7
Analytical Procedures

Figure 6 presents two analytical procedures used in our institute for the determination of trace elements in rainwater samples. Before 1993 the electrochemistry was the method of choice because of its high sensitivity. However, only a limited number of elements can be determined by this method – a major disadvantage of the electroanalytical techniques. Since 1993, new analytical methods, such as potentiometric stripping analysis, ion chromatography [14], ICP-OES [19] and ICP-MS have been used. With these methods it is possible to determine a significantly higher number of pollutants in the rain samples.

3.8
Results and Discussion

The conductivity of rain samples is closely connected with the concentration of H^+ ions. Only in sampling areas where there is sea spray, can the high concentration of sodium chloride influence the conductivity. Figure 7 shows the relative frequency histogram of pH values in a sampling area close to a highway (Fig. 3.7a) and in a rural region of Germany (Fig. 3.7b). A few years ago, low pH values in rainwater were observed, but due to the changes in the automobile technology and more rigorous control of industrial pollution, the atmospheric input of acid-producing gases has been reduced; as a result the pH of rainwater changed to higher values. This positive trend is also observed in other sampling areas.

Figure 8 presents the data for the long-term monitoring of cadmium and lead in rainwater collected in different sampling areas. A few years ago, in areas with high industrial pollution not only the concentration of lead but also that of cadmium in rain was significantly higher than in the city or in rural areas.

Due to contamination control by the use of modern technologies and more efficient filters, the concentration of these elements in rainwater has decreased. In

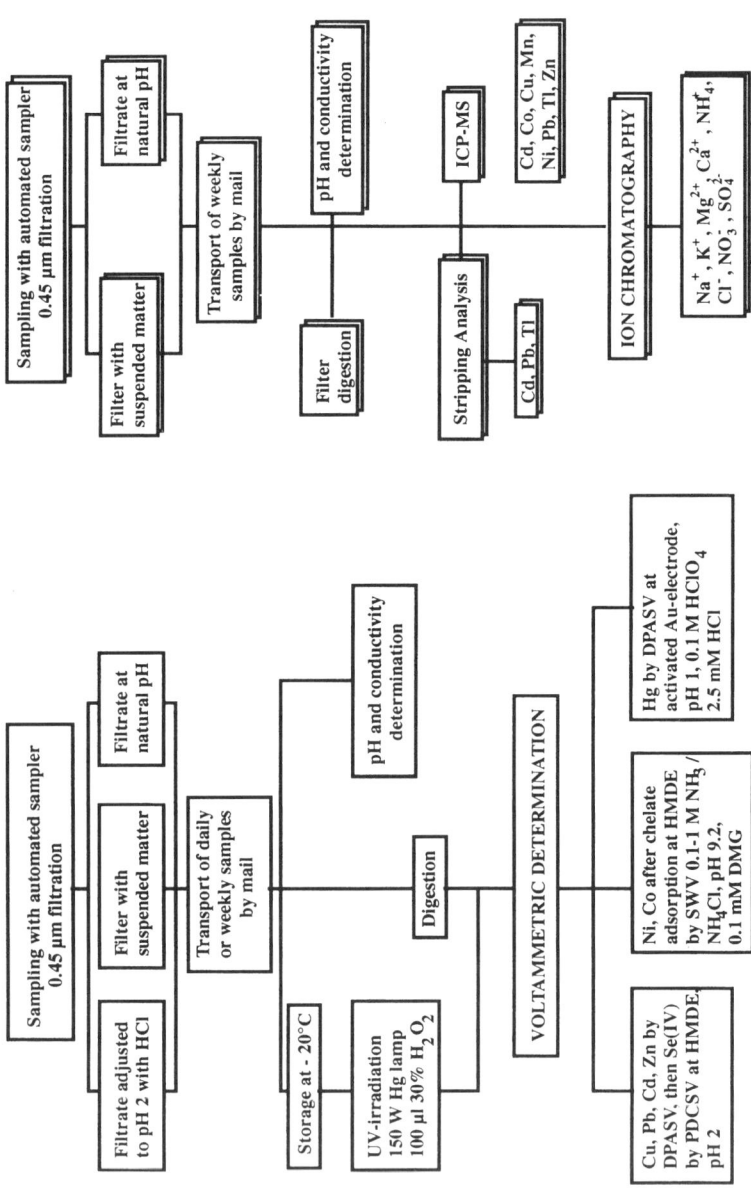

Fig. 3.6. Procedures used for the determination of trace elements (ions) in rain water samples by electroanalytical and other instrumental techniques

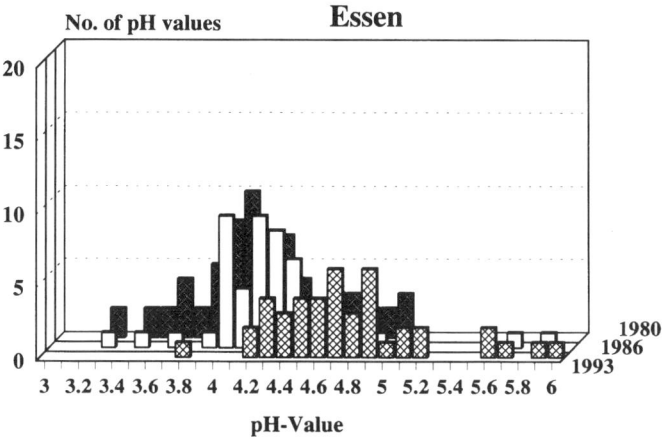

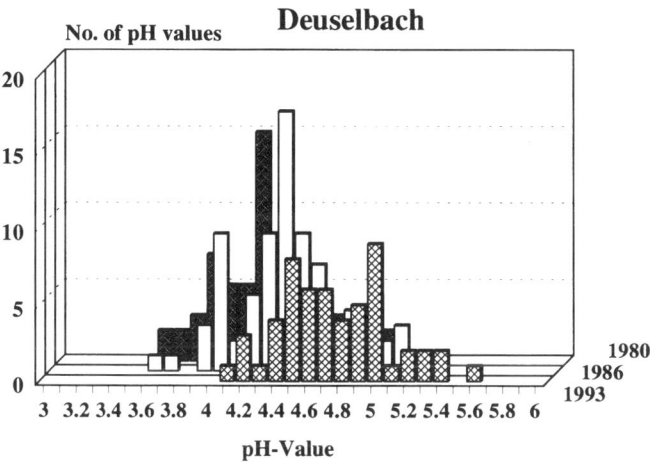

Fig. 3.7. Distribution of long-term changes in the pH values in a rural (Deuselbach), and an urban region (Essen)

the city areas the concentration of related to traffic density. The reduction of the lead concentration in the fuel has also reduced the lead concentration in rainwater. The concentration of cadmium is related to the energy and heating systems. New technologies in building have reduced energy consumption, leading to a lower concentration of cadmium in rain. In rural regions low concentrations of both elements are observed. Minimizing the anthropogenic contamination sources also reduces the concentration of cadmium and lead in areas without any human activities. Because of the more or less stable situation in the economy and the efforts made to protect the environment, the pollutant concentration in rainwater will be stabilized towards low levels in the next few years.

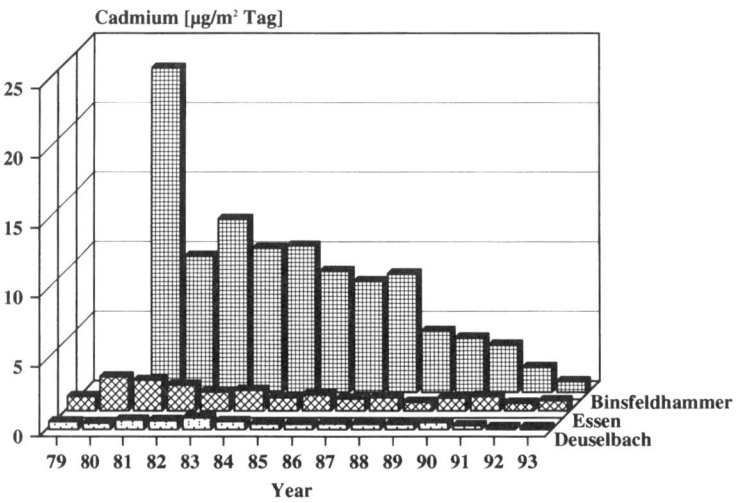
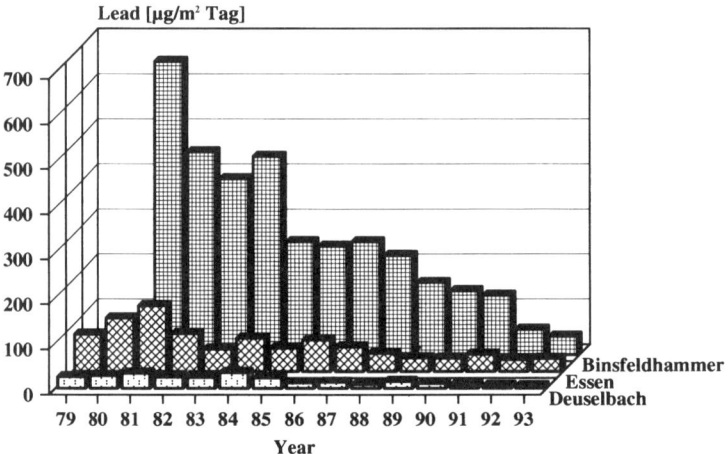

Fig. 3.8. Long-term deposition of cadmium and lead in a rural (Deuselbach), an urban (Essen), and an industrial region (Binsfeldhammer)

3.9
Conclusions

From the available experimental data it can be concluded that there is still no reliable method avoilable for the objective and absolute determination of pollutants in precipitation. Information obtained by one sampler along with its method of determination are not comparable with another sampler using a different method of determination; only small data sets can be compared.

New collection techniques have to be developed in order to quantify amounts of rain [17] and snow. The reproducibility and comparability of sampling techniques must be improved. A combination of analytical procedures together with reproducible sampling equipment can minimize the large variations in pollutant concentrations in rain samples that have been measured in different laboratories around the world.

3.10
References

1. Berner EK (1992) Encyclopedia of Earth System Science, Vol 1. Academic Press, p 437
2. Freedman B (1992) Encyclopedia of Earth System Science, Vol 1. Academic Press, p 1
3. VDI-Richtlinie 3786, Blatt 7
4. Nürnberg HW, Valenta P, Nguyen VD (1983) Proc. Int. Conf. Heavy Metals in the Environment. Springer, Berlin Heidelberg New York, Vol 1, p 70
5. Winkler P, Jobst S, Harder C (1989) PTP Bericht der GSF München, ISNN 0176-0777
6. Winkler P, Schulz M, Dannecker W (1991) Fresenius J Anal Chem 340:575
7. Gravenhorst GC, Persecke C, Rohbock E (1980) BMFT-Abschlußbericht Nr. 10402600
8. Lareen DPM, Harrison RM (1981) Anal Chem 53:345–350
9. Peden ME, Skowron LM (1978) Atmos Environ 12:2343–2349
10. Wagner F (1989) Ber. KFA Jülich, Jül-2290, D82 (Diss. T.H. Aachen)
11. Wagner F, Valenta P, Nürnberg HW (1985) Fresenius Z Anal Chem 320:470–476
12. Boutron CF, Hong S, Candelone JP (1995) Proc Int Conf "Heavy Metals in the Environment", Hamburg September 1995, pp 28–33
13. Rojas CM, Injuk J, Van Grieken RE (1993) Atmos Environ 27A:251–259
14. Zlotorzynska-Dabek E, Dlouhy JF (1994) Journ Chromatog A 671:389–395
15. Lum KR, Kokotich EA (1987) Sci Total Environ 63:161–173
16. Slanina J, Möls JJ, Baard JH, van der Sloot HA, van Raaphorst JG (1979) Intern J Environ Anal Chem 7:161–176
17. Bächmann K, Steeg K-H, Groh T, Röder A, Haumann I, Boden J (1992) Intern J Environ Anal Chem 49:87–101
18. Hofmann H, Hoffmann P, Lieser KH (1991) Fresenius J Anal Chem 340:591–597
19. Boutron CF, Görlach U (1990) NATO ASI Series Vol. G 23:137 151, Metal Speciation in the Environment, ed. Broekaert JAC, Gücer S, Adams F, Springer Verlag, Berlin
20. Kerkes J, Freedmann B (1988) Arch Environ Contam Toxicol 18:183–192
21. Rajsic S, Otten P, Van Grieken R (1991) Environmental Technology 12:257–261

Chapter 4
Sampling of Sea and Fresh Water for the Analysis of Trace Elements

ECKARD HELMERS

4.1 Introduction

The recognition of the importance and influence of the sampling and decomposition steps was one of the main advances of trace element research and analysis in the last two decades.

Contamination during sampling and decomposition may falsify analytical results by several orders of magnitude. By comparison, errors in the instrumental analysis step are normally much smaller and – if the analysis can be repeated – reversible. This fact seems to be well established within the earth sciences since high quality data are a precondition for the quantification and understanding of elemental pathways between atmosphere, hydrosphere, lithosphere and biosphere. In contrast, for example within the medical sciences, the problem of contamination during sampling had been partly neglected until recently. Risks of contamination can be controlled by technical expenditure and methodical tricks.

In the case of sea water sampling the risk of contamination is particularly high. Thus we use it here as an example for the technical and methodical solutions available for investigations in the aquatic environment. Technical and scientific progress is best illustrated by the history of lead concentrations already published: nearly all data prior to 1970 exceeded natural lead concentrations by up to four orders of magnitude as a consequence of contamination (Fig. 4.1).

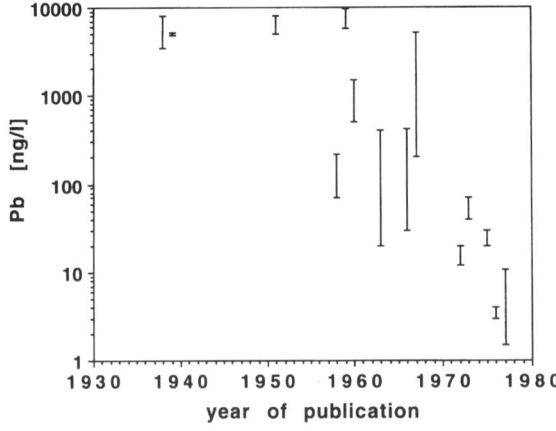

Fig. 4.1. The "contaminated history": sea-water Pb concentrations published between 1938 and 1980 (after Mart, 1979a and Schaule, 1974)

The lower the natural concentration of an element and the larger the extent of the anthropogenic uses of the element, the higher is the risk of contamination during sampling and analysis. Typical elements for which the risk of contamination is high are Al, Fe, Pb and Zn. On the other hand, not only the analysis of ubiquitous metals, but also those used anthropogenically may suffer from contamination: More than 70 years ago, it was believed that the sea contained gold in the milligram-per-liter-range. Persuing this belief, Fritz Haber studied the distribution of Au in sea water measuring concentrations between 1 and 8000 ng/l (Haber, 1928). Obviously the golden ring on one of his fingers had contaminated the samples because the actual concentration of Au in sea water is around 40 pg/l (Koide et al., 1986).

Representativeness is another topic connected with the sampling strategy. Since the time scales of dynamic processes are very different and strongly related to the individual water matrix this will only be briefly reviewed here.

4.2
Sample Handling

4.2.1
Preparatory Steps

4.2.1.1
Clean room Requirements and Behaviour of Personnel

There are numerous sources of contamination: obvious ones such as contaminated sampling bottles, samplers or the contact with hands or acids; less obvious ones such as air particles. These are most critical during the various steps of ultratrace analysis (< 1 µg/l): cleaning procedure, handling of samplers before and after filling, acidification, aliquotation, and transport (Mart, 1982). Every time a sample is open, it must be protected from dust particles. Human beings can be regarded as "particle generators". Consequently the personnel must be well trained, motivated and able to think critically during every stage of sample handling. Every step must be checked for the risk of contamination: could there have been abrasion of the wires, was there a clean support for the lids, was the winding touched? Hands or clothing must never be held above the open sample. The wearing of clean gloves (for example: polyethylene PE) is essential, and the gloves should not be powdered (possibly Zn powder). The clothing must be suited to the purpose (no wool), and smoking prohibited. In the ultratrace contamination range cleaned (filtered) air is normally required; that means clean-bench working, especially during sample processing in a technical environment (e.g. on a ship). The style of working must be specially adapted to clean-bench conditions: a clean air stream must always pass over open samples in the direction of the operator.

In order to allow the whole sample handling and analysis of sew water within a clean atmosphere, a clean room laboratory container was specially developed for shipboard use (Helmers, 1991). This container (size: $20 \times 8 \times 8$ feet, manufacturer Seus Corp. Wilhelmshaven, Germany) included an entrance room for changing clothes and sample storing, a working room (US class 1000), a clean water supply,

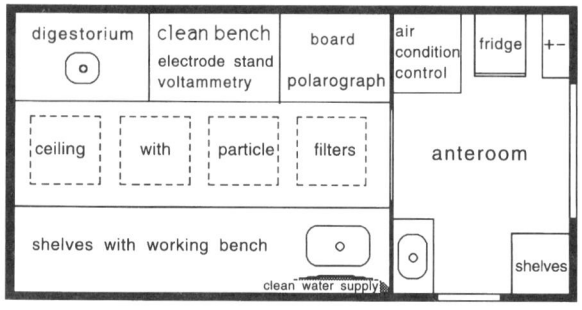

Fig. 4.2. Newly constructed clean room container laboratory for shipboard use (Helmers, 1991)

a separate clean bench (US class 100) and a digestorium for clean acid destillation (Fig. 4.2).

4.2.1.2
Selection of Labware and Sampling Bottles: Cleaning Procedures

During sampling and analysis of trace elements, metallic materials must be avoided as far as possible. All labware including scissors and tweezers must consist of plastic material. Commercially available laboratory equipment, for example clean benches or working desks, must be modified to hide or eliminate metallic parts. Since acids are widely used to stabilize the samples, metallic parts inside the laboratory would corrode and contaminate the laboratory atmosphere. One of the most important decisions is the selection of bottle material for sampling and storing. There is a variety of suitable materials available, the three most common ones are listed in Table 4.1. PE, Teflon and quartz materials are suitable for non-contaminat-

Table 4.1. Materials suitable for sampling and storing natural waters for subsequent analysis of trace elements

Material	Advantage	Disadvantage
Polyethylene PE (LDPE "Kautex")	Very low price, simple handling, reliable low blank values because of one-way use	Not for Hg analysis
Teflon (several types)	Stable (multiple use) teflon/titanium sampler commercially available	Expensive, not for Hg analysis, high weight (block teflon), transparent teflon fragile, new material possibly contaminated, memory effects possible
Quartz glass	Low blank values even after multiple use	Expensive, fragile

ing sampling if they are adequately precleaned and used. My own experiences have shown that bottles which are used several times require a high effort in re-cleaning involving elevated risks of contamination (see below). For example, the steaming-equipment for Teflon bottles recommended in the literature may not reliably clean the whole inner surface of the bottle, the winding, the lids. Generally, steaming devices (Tschöpel et al., 1980) can cause severe corrosion within the laboratory especially if they contain metallic fittings. PE-bottles, in contrast, can be effectively cleaned using hot baths with dilute acids: A barrel also made of PE can hold a large number of PE bottles (Fig. 4.3). The acidified water (hydrochloric or nitric acid) is heated indirectly by an Erlenmeyer flask containing the heater. A magnetic stirrer is used for mixing (modified from Mart, 1982). After heating in the acid bath, bottles and lids are rinsed with ultraclean water and sealed in new PE bags. If necessary, conditioning can be repeated in a second barrel to further improve blank values. The whole procedure is given in Table 4.2. Since these bottles are used only once, there is no danger of contamination from previous use.

Applying this cleaning method, around 300 samples of open ocean surface seawater were collected into PE bottles ("Kautex") allowing detection of natural Pb concentrations between 1 and 50 ng/l without a single outliner (Helmers and Rutgers van der Loeff, 1993). Blank values of the so-cleaned PE bottles are below the voltammetric detection limits of 0.1 ng/l (Cd, Pb), and 1 ng (Cu), respectively.

4.2.1.3
Purification of Water and Acids

The water for rinsing, conditioning and cleaning processes must be generated by an ion-exchange facility (18-MΩ resistance of outflow), such as are commercially available from a number of producers. Water distillation devices will not supply either the quality or the quantity necessary. An ion-exchange facility can be easily operated on board a ship or another mobile carrier. Acids used for cleaning processes (Table 4.2) must be of commercially available p.a.-quality or better. Acids used for the conditioning of the sampling bottles and for the acidification of the samples must be of maximum purity, i.e. commercially available suprapure® or subboiled. A subboiling distillery (Tschöpel et al., 1980) operating either with

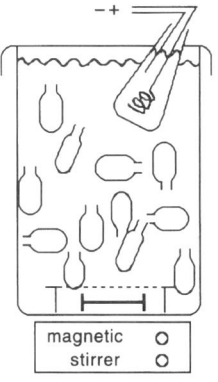

Erlenmeyer flask with heater and thermometer

50-liter-barrel (PE) with acidified clean water (60 °C)

stirring

Fig. 4.3. Simple and effective cleaning device for PE sampling bottles

Table 4.2. Cleaning and conditioning scheme for PE sampling bottles

Precleaning and degreasing in a laboratory disher
⇓
Rinsing with ultraclean water
⇓
Precleaning in an acidified water bath (approx. pH 0.5)
at room temperature for one week
⇓
Rinsing with ultraclean water
⇓
Heating in an acidified water bath (Fig. 4.3) at approx. pH 1
for one week at 60 °C (diluted acid afterwards rejected)
⇓   ↑
Rinsing with ultraclean water
⇓
Filling with ultraclean water, acidification (10 ml per 500 ml of volume) with conc. nitric acid (subboiled), enveloped two-fold in PE-bags

hydrochloric or nitric acid can be operated within the clean area on board a ship (Fig. 4.2).

4.2.2
Contamination Effects: an Example

During the author's investigations, the appearance of contamination and ways of overcoming it were intensively studied. In Fig. 4.4 results of the Pb-analysis of surface seawater, collected on two expeditions from Cape Town to Bremerhaven through the eastern Atlantic Ocean, are displayed. Samples were taken by the same sampling procedure (see below) on both cruises and analyzed by voltammetry (Helmers et al., 1991). However, on the first cruise, expensive Teflon bottles were used that had been cleaned by a steaming device (Sect. 4.2.1.2) and conditioned with acidified ultraclean water. Although this is a common cleaning method, the Teflon bottles were nevertheless the main source of contamination here (for details, see Sect. 4.2.1.2). On the second cruise PE bottles were used, cleaned as recommended in Sect 4.2.1.2. From these results contamination effects can be described as follows:

- contamination is irreproducible and erratic;
- background contamination (in this case approx. 5 – 10 ng/l) prevents the resolution of events below this range of concentration;
- single outliners can exceed the natural metal concentration by up to several orders of magnitude.

Nevertheless, even with insufficient contamination control, as on the first cruise (Fig. 4.4), a few insights can be gained: since generally double samples were taken on the first cruise, some outliners could be identified afterwards. The increase in the concentration of lead in the North Atlantic between 30 and 50° N was observed. However, the important signal that was registered later in tropical Atlantic

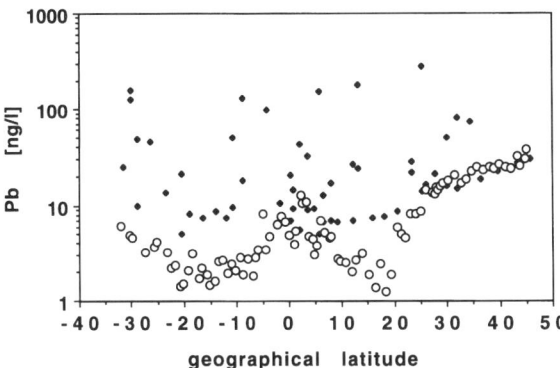

Fig. 4.4. Illustration of contamination: surface sea water sampled in unclean Teflon bottles (filled points, 1989) and in clean PE bottles (open circels, 1990) on two cruises of the RV "Polarstern" within the same area of the eastern Atlantic Ocean (geographical longitude: 10° E to 30° W)

(Helmers and Rutgers van der Loeff, 1993) remained invisible in 1989 because of the background contamination.

Concerning other trace elements, such as Al, Cd and Cu, contamination effects in the same samples were much lower or missing. Consequently, Pb analysis reveals an unsuccessful contamination control in a highly sensitive way.

4.2.3
Need for Filtration and Filtration Design

Natural waters contain suspended particulate matter (SPM) in concentrations varying by several orders of magnitude (Table 4.3). With the exception of open ocean waters, the filtration of natural waters is generally required since the SPM can contain large amounts of certain elements. The portion of metals associated with SPM can easily exceed amounts contained in the dissolved phase (Sect. 4.5). Without a filtration step, acidification would leach elements from the SPM in the sample. Since this step is hardly reproducible and, moreover, particulate freights in natural waters can be highly variable over time (Guhr and Weber, 1994), this would lead to a set of highly scattered data unsuitable for interpretation or trend detection. Concentrations of elements in the SPM can be given on the basis of either volume or weight (Sect. 4.5); the latter characterizing the nature (biogenic,

Table 4.3. Concentrations of suspended particulate matter [SPM in mg/l] in sea-surface and fresh waters

Area	Minimum	Maximum
Open ocean (Atlantic[a])	< 0.001	0.1
Shelf sea/coastal (North Sea[b])	0.1	10
Estuaries (River Weser[c])	10	200
Rivers[d])	0.5	> 4000

References: [a] Helmers, 1991; [b] Nolting and Eisma, 1988; [c] Duinker et al., 1982; [d] Schwoerbel, 1993.

geogenic or anthropogenic) of the investigated SPM. Since the weighing of particulates is the limiting step with respect to sensitivity, the data from offshore waters is limited.

Filtration can serve two purposes: separation of the SPM in order to receive a "clean" water sample (the SPM is rejected) or enrichment of the SPM for a subsequent separate investigation. For the first purpose, a water sample should be filtered under pressure (clean nitrogen) through an acid-precleaned plastic filtration unit. For example, filtration units made of polycarbonate are available from Sartorius Corp., adapted to 47 mm filter size. Suitable filters made of polycarbonate are available e.g. from Nuclepore Corp. (Type SN 11107). These filters possess a low risk of contamination if they are precleaned in highly-diluted ultraclean acids. The filtration of natural waters is illustrated schematically in Table 4.4. The procedure may be simplified for the filtration of fresh waters. After filtration, the filter can be rejected or introduced into a decomposition step together with enriched SPM (Sect. 4.4). The filtration and handling of filters must be carefully performed in a clean atmosphere (e.g. clean bench). Following an international convention, filtration through a pore diameter of 0.45 µm is differentiating between dissolved and particulate phases.

In open-ocean seawater sampling, filtration is usually avoided because the SPM concentration is very low and because the filtration process itself generally carries a contamination risk. However, recent research revealed that the former is not valid for all elements. For example, even in the open-ocean surface, concentrations of

Table 4.4. Flow scheme for the pressure filtration of natural waters

---
Mounting of the (precleaned) filter (e. g. polycarbonate)
in the opened and precleaned filtration unit
⇓
Careful closing of the filtration unit
(danger of filter damage)
⇓
Filtration of 100 ml ultraclean water
⇓
Conditioning of the filter (only for sea-water sampling):
filtration of 100 ml unpolluted sea water (rejected)
⇓
Conditioning of sampling bottle:
filtration of 500 ml water (rejected)
⇓
Sampling: filtration of 500 ml water (later acidified)
⇓
Sampling of SPM from seawater: filtration of 5×1 ml ultraclean water
to remove sea salt; blow-drying of SPM with nitrogen
⇓
Removing of loaded filter, packing in precleaned containers
⇓
Rinsing of filtration unit with acidified ultraclean water,
then with neutral ultraclean water

particulate Fe cannot be neglected in comparison with the amounts of dissolved Fe (Helmers, 1991; data given in Sect. 4.5).

If the SPM from offshore waters is the topic of interest, larger volumes of water have to be extracted. Pressure filtration is not suitable here because colloid material will quickly block the filter pores. In this case, only a continuous-flow extraction device – for example a centrifuge – allows larger amounts of particulates to be sampled. After filtration or extraction of the sea water, the resulting particulates must be carefully washed with ultraclean water in order to separate them from remaining sea salt; large amounts of sea salt could interfere with a subsequent analysis by atomic spectrometry.

To guarantee the non-contaminating accumulation of SPM from sea water for subsequent analytical investigations (e.g. weighing, AAS and POC-analysis), and to exclude a loss of particulate material during handling, a processing scheme was developed (Fig. 4.5). The rotor of a continuous-flow centrifuge must be made of titanium, which does not cause contamination. It can be equipped either with a

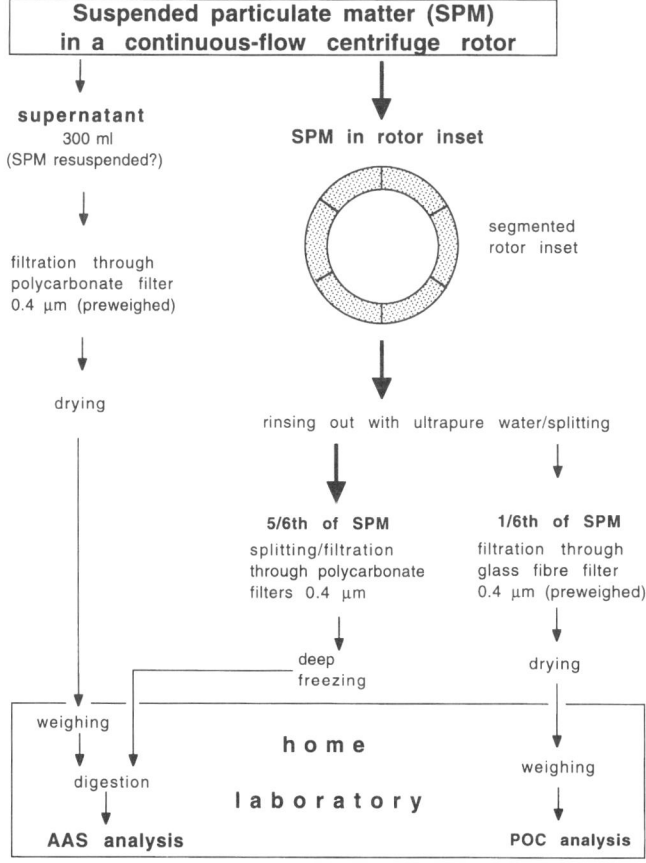

Fig. 4.5. The SPM sample treatment and processing protocoll

polycarbonate inset (Fig. 4.5) or with Teflon foils (Schüßler and Kremling, 1993). A rotational acceleration of over 18 000 xg is required for an effective extraction of particulates. The smallest diameter of particles being retained this way is in the order of 0.1 µm (Schüßler and Kremling, 1993).

Correct weighing of the SPM-loaded filters in the home laboratory is perhaps the most difficult analytical problem. Even when using a high precision laboratory device for the weighing, the uncertainty was at least ± 0.3 mg. Moreover, since weighing itself is also a possible source of contamination, the SPM may be aliquoted. Only aliquots needed for less-sensitive analysis parameters (such as particulate organic carbon POC) may be weighed (Fig. 4.5).

4.2.4
Storage

Most authors recommend acidification of water samples immediately after sampling, independent of storage temperature. Hydrochloric acid should be used if the samples are to be analyzed later by voltammetry whereas nitric acid is mostly recommended for later spectroscopic analysis (amount: approx. 1 µl ultraclean acid per ml sample). If samples are stored unacidified and unfrozen, adsorption of elements on to the walls of the vessels will take place, thus lowering the original concentrations of dissolved elements (Tschöpel et al., 1980). A later acidification may partly redissolve elements. However, the rate of this reaction may vary according to the individual sample composition. Samples may be stored unacidified but frozen in order to preserve the original distribution of the species. On the other hand, even metal compounds with humic substances are stable down to pH 2.3 (Helmers, 1994).

It is advantageous to check for contamination immediately after sampling in order that possible sources of contamination are identified *during* the expedition rather than when the sampling stage is over. This check may be performed by voltammetry on the ship or sampling vessel. Later, in the home laboratory, the elemental concentration may be partly measured again (Fig. 4.6). Sea water samples acidified with hydrochloric acid down to pH 2.3 and stored at 10 °C were found to show the same Pb concentrations (1 – 10 ng/l) over at least two years of storage in PE bottles.

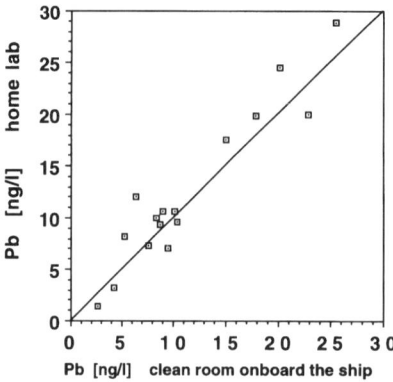

Fig. 4.6. Contamination and stability control: comparison of analytical data from identical samples obtained on board a ship and in the home laboratory several weeks later (open-ocean surface seawater)

To prevent any bacterial activity during storage, SPM-loaded filters may be stored in precleaned petri dishes (Polystyrene) either dry in a desiccator or deep frozen.

4.3
Sampling Procedures

4.3.1
Collection of Sea Water

The particular problems of sea water sampling and analysis were mentioned in the introduction: the very low concentration ranges and the need to sample from a floating vessel that contaminates the surrounding water. Over the past two decades, much effort has been made to improve the sampling techniques. At first, ultraclean deep water samplers were developed, enabling the typical distribution pattern of elements, such as Pb, within the water column to be the determined (Schaule and Patterson, 1980).

4.3.1.1
Water Column

Suitable deep water samplers must be moved unopened through the contamination cloud surrounding the research vessel. Commonly, "close-open-close"-samplers of General Oceanics Corp. are used. The larger the sampler volume (up to 30 liters), the smaller is the risk of contamination. These samplers are handled with a Hostalen-coated wire and are opened at approx. 10 m depth by hydrostatic pressure. Samples from a depth of several thousand meters may be obtained. On board, the samplers must be carefully cleaned and handled. New samplers or samplers which have not been in use for a long period are appropriately cleaned by passing through very deep water. Various sampler techniques have been reviewed by Schaule and Patterson (1980).

4.3.1.2
Surface Sea Water

20 years ago, surface seawater was sampled from a rubber boat after leaving the research vessel. Several factors such as wind and current direction must be considered when choosing an unpolluted sampling area far from the vessel. Nevertheless, a specially-constructed and precleaned sampling gear made of plastic materials is necessary for sampling from the rubber boat (Mart, 1979b).

More recently specially designed samplers were applied to allow surface water to be sampled directly from the vessel. For example, a "Mercos"-sampler (Freimann et al., 1983), consisting of a Teflon body and titanium fittings (Hydrobios Corp., Kiel, Germany), was adapted to PE sampling bottles by the author (Helmers, 1991), since the original Teflon bottles had been a source of contamination (Sect. 4.2.2).

In this way, Al and Pb were analyzed at ultratrace levels throughout the surface of the Atlantic Ocean (Helmers and Rutgers van der Loeff, 1993):

The sampler was installed at the end of the ship's bow boom. This position is 14 m above the water surface and 8 m in front of the ship. The sampler was handled with a high-grade steel winch, a 5 mm Kevlar line and a counterweight covered with synthetic polymer material, and operated in about 1–2 m of water depth.

Special care must be taken to avoid contamination from the ship: it is reducing the speed down to 1–2 knots and then turning its front towards the wind. Immediately after the ship stops, the closed sampler is submerged, opened by a falling weight and allowed to fill with seawater. Commercially available falling weights are filled with lead. Since this may be a contamination source for Pb, new weights with an iron core were prepared. The distance between the ship's bow and the sampling position below the bow boom must be long enough to prevent contamination from spray water of the ship's bow.

The next development was the sampling via a continuously-pumping snorkel system (Kremling, 1985): an approx. 2-m long snorkel made of stainless steel (diameter. 8 cm, wall thickness: 0.9 cm) is attached to the lid in the hydrographical moon pool. Seawater is continuously transported by means of a Teflon pump (e.g. Almatec Corp., flow rates 10–70 l/min.) through a PE tube, which is fixed at the end of the snorkel to a Teflon block with a hole. A water tap of this seawater supply line was installed in a clean bench where samples were filled into PE-bottles (Fig. 4.7). The snorkel must always be long enough to reach a non-turbulent water-zone below the ship.

The advantage of sampling by a continuously-pumping snorkel system is that samples can be collected on the way. It offers the possibility of high-resolution sampling of the ocean surface, enabling even small-scale variations to be recognized.

The number of samples to be taken is limited only by the subsequent analytical capacity. The frequency of sampling is variable and may be adapted to the oceanographic situation during an expedition.

4.3.2
Collection of Fresh Water: Lakes, Rivers, Estuaries

During the collection of fresh water or estuarine water from large rivers similar sampling methods are applied as for sea water sampling. For example, a yacht can be equipped with a telescope arm that collects water by means of a non-contaminating tubing system. In this way, lateral and longitudinal distribution patterns within a stream may be investigated. Generally, the contaminated water area around the ship must be avoided. Deep water samples of rivers and lakes can be obtained by suitable samplers, as described in Sect. 4.3.1.1. Samplers and wires must consist of only a minimum or no metallic materials. Generally, there are two technical principles: samplers for different depths are either mounted along a wire and closed in the water column by falling weights, or they are mounted side by side ("rosette") and operated by remote control. Surface sampling in smaller lakes may be performed from the bank using a precleaned sampling gear. The frequency of sampling must be adapted to the individual time scale. For example, in an estuary, elemental parameters may dramatically change within a few hours. Examples of specially-de-

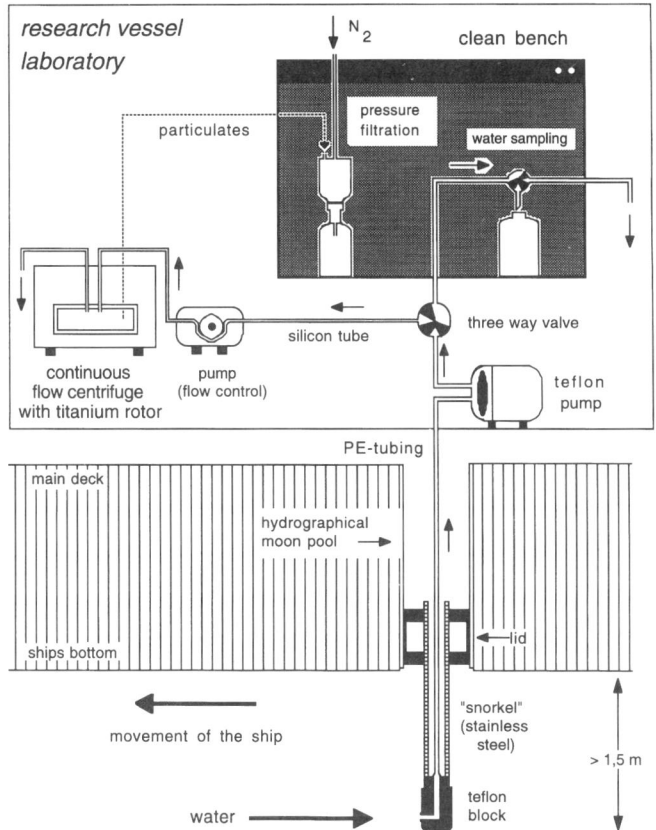

Fig. 4.7. Design of SPM and seawater sampling equipment used on board a research vessel

signed sampling strategies are given in the literature, e.g. for lakes (Klapper et al., 1994), rivers (Guhr and Weber, 1994), wastewater (Dick, 1994) and even groundwater (Puls, 1994).

4.4
Digestion

4.4.1
Digestion of (Filtered) Water

A filtered water sample may be analyzed either directly by voltammetry or by spectrometry, the latter method requiring a matrix-separation step in the case of sea water (see Sect. 4.5). Although voltammetric analysis can in most cases be performed immediately after acidification, higher concentrations of dissolved organic matter (DOM) or very low element concentrations (1–10 ng/l) require photo-digestion of the water sample before analysis (Dorten et al., 1984). During the voltammetric analysis of seawater, UV-irradiation increases the detection power because the signals are sharpened (Helmers et al., 1991). Also, time-resolved irradiation experiments can supply information on the nature and distribution of different species of metals (Helmers, 1994).

Mart (1980) suggested a simple, self-designed UV-irradiation device adapted to voltammetric analysis cells in order to prevent contamination during the irradiation procedure. A 150 W mercury vapor lamp is sufficient for low-DOM waters (such as seawater or freshwater); it also anables kinetic irradiation experiments to be performed. Irradiation can be performed either with or without oxidizing agents (preferably H_2O_2), which must be of maximum purity (e.g. suprapure®).

4.4.2
Digestion of Particulates

Before enriched SPM can be submitted to elemental analysis (Fig. 4.5), it must pass through a mineralization step. There are several technical approaches, which are reviewed in detail in this book. The pressurized decomposition with ultraclean nitric acid in Teflon bombs (e.g. from Berghof Corp., Germany) at 200 °C is sufficient for SPM that mainly consists of biogenic material (algae), as extracted from open-ocean seawater (Helmers, 1991). SPM with a higher geogenic share may require HF-assisted decomposition in order to dissolve the silicate.

4.5
Typical Concentration Levels of Selected Trace Elements in the Aquatic Environment and Suitable Analytical Methods

Table 4.5 lists the concentrations of some elements in the aquatic environment, giving an overview of magnitudes. The elements Al, Cd, Cu, Fe, Mn and Pb are among the most-studied metals in the environment. In the hydrosphere, Cd, Cu and Pb are contained in the dissolved state at trace or ultratrace (< 1 µg/l) levels only; they are frequently investigated in order to detect anthropogenic input. Al, Fe and Mn in waters are important geogenic tracers. The concentration levels reviewed in the table do not reflect all possible variations in worldwide waters but they are representative. More data are offered in specialized review publications, e.g. Whitfield and Turner (1987) on sea water.

However, the data in Table 4.5 clearly confirm the importance of filtration during the investigation of shelf sea water (< 200 m of sea depth), river water and partly even open-ocean sea water (Al, Fe).

Analysis of elements dissolved in sea water is most difficult because salinity interferes with spectrometry. A matrix separation (extraction) step (Kremling, 1983) is therefore required prior to analysis by graphite-furnace atomic absorption spectrometry (GF-AAS) or total reflection X-ray spectrometry (TXRF; Haarich et al., 1993). Voltammetric analysis (frequently adapted for Cd, Cu and Pb investigations in sea water) does not require matrix separation. Although in many cases several analytical methods are available for each element, there are some element-

Sampling of Sea and Freshwater for the Analysis of Trace Elements

Table 4.5. Surface concentration levels of selected trace elements in the aquatic environment and suitable analytical methods

Element/matrix		Dissolved (μg/l)	SPM (μg/l)	SPM (μg/g dry weight)	Analytical methods (dissolved elements)
Al	Open ocean	0.007–2.1	0.002–0.2	300–2×10^3	Fluorimetry; extraction/GF-AAS
	Shelf sea	1–4	5–14	8×10^3–2×10^4	Fluorimetry; extraction/GF-AAS
	Rivers[b]	1–15	5–50		AAS, ICP
Cd	Open ocean	0.001–0.03	5×10^{-6}–0.001	1–8	Voltammetry; extraction/GF-AAS
	Shelf sea	0.005–0.02	0.004–0.05	4–100	Voltammetry; extraction/GF-AAS
	Rivers	0.05–0.6	0.2–0.4	3–10	GF-AAS, ICP
Cu	Open ocean	0.03–0.2	3×10^{-4}–0.002	3–21	Voltammetry; extraction/GF-AAS
	Shelf sea	0.2–0.5	0.005–0.01	8–13	Voltammetry; extraction/GF-AAS/TXRF
	Rivers	2–10	3–7	60–90	GF-AAS, ICP
Fe	Open ocean	0.004	0.001–0.1	20–1000	Extraction/GF-AAS
	Shelf sea	0.1–10[c]	4–9	4×10^3–2×10^4	Extraction/GF-AAS/TXRF
	Rivers	10–30	2×10^3–6×10^3	4×10^4–5×10^4	AAS, ICP
Mn	Open ocean	0.06–0.3	0.0001–0.005	5–130	Extraction/GF-AAS
	Shelf sea	0.3–1.4[a]	0.03–0.07	30–80	Extraction/GF-AAS/TXRF
	Rivers	10–70	100–300	2×10^3–4×10^3	AAS, ICP
Pb	Open ocean	0.001–0.03	2×10^{-5}–3×10^{-4}	0.8–10	Voltammetry; extraction/GF-AAS
	Shelf sea	0.01–0.05	5×10^{-3}–0.01	6–20	Voltammetry; extraction/GF-AAS/TXRF
	Rivers	0.01–10		20–800	GF-AAS, ICP

Notes: Open ocean (Atlantic) and shelf (North Sea) data from Helmers, 1991 and Helmers and Schrems, 1995 ([a]Kremling, 1985); river data (River Weser) from Duinker et al., 1982 ([b]Stabel et al., 1991; [c]estimated from various sources). SPM = suspended particulate matter. Further abbreviations explained in Sect. 4.5.

specific peculiarities: for example, Al can be elegantly analyzed in high-salinity waters by fluorimetry (Hydes and Liss, 1976); Pb seems to be detected most sensitively in sea water by voltammetry (Helmers and Rutgers van der Loeff, 1993); and ICP-MS (inductively-coupled plasma with mass spectrometric detection) is a powerful tool for analyzing heavy elements in fresh waters.

After acid decomposition, elemental concentrations of SPM can be quickly and reliably analyzed by GF-AAS.

4.6
Quality Assurance During Analysis and Data Evaluation

4.6.1
General Aspects

Certified reference materials should be analyzed regularly during an analytical project. A number of fresh and high salinity waters are commercially available. Standard reference sediment material is suitable for supporting SPM analysis. Blank values of labware (bottles, filters, etc.) and processes must be determined regularly. Repeated measurements provide information on analytical and sample stability (Chapter 2.4).

Apart from these analytical steps, the results themselves reveal information on reliability: How wide is the range of results in an equilibrium sampling area? Do the results reflect trends or do they only show a cloud of points? If there are trends, will they be reproducible during the next sampling? A reproducible detection of a trend itself is important evidence – but not a sufficient proof – for successful quality management. The discovery of a trend, if there is one, should principally be an aim of an analytical project.

4.6.2
Trend Monitoring: Decrease/Increase Verification

Specimen banking offers an opportunity of reliably recognizing analytical trends. Since water samples are normally not included in environmental banking programs, assessing a change in data within time is difficult. Figure 4.1 clearly elucidates the problem: the decrease in marine Pb concentrations reported in the literature was solely a result of methodical improvement. In reality, Pb concentrations should have increased during this period. However, in the 1980s, analytical results revealed that the concentrations of Pb in the Northeast Atlantic Ocean had halved (Fig. 4.8). The verification of this trend is mentioned here as an example:

- The analytical method used was the same during the whole time: voltammetry with a rotating mercury film electrode, offering the possibility of zero-blank-working, i.e. blank below detection limit.
- The samples were taken within in same geographical area.
- The sampling was performed from south to north, i.e. from the area of low concentrations to higher concentrations (no danger of contamination from the higher polluted area).

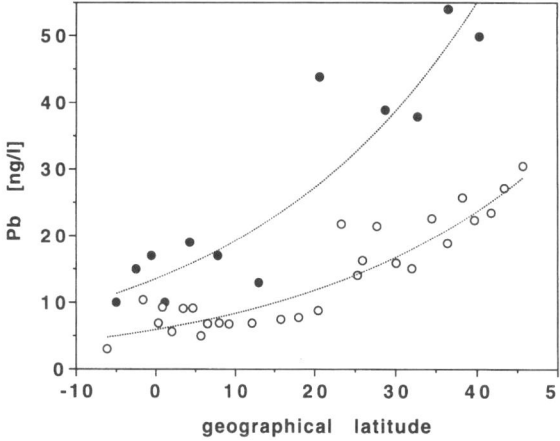

Fig. 4.8. Surface lead concentrations in the open Northeast Atlantic Ocean (Helmers et al., 1991): comparison of the results from expeditions in 1981 (closed circles) and 1989 (open circles). Geographical longitude: 5° W to 30° W

- Both expeditions (1981 and 1989) measured a comparable "background" concentration level in the tropical Atlantic Ocean (around the equator).
- The detected trend was congruent with the results of other environmental investigations.

However, as stated in Sect. 4.2.2, improved sampling techniques resulted in a lower background Pb concentration (down to 1 ng/l) in the subtropical Atlantic (around 17° N). Nevertheless, the assessment and modeling of the trend (Helmers et al., 1990) in the polluted area (> 25° N) have been confirmed.

4.7
References

1. Dick EM (1994) Water and wastewater sampling for environmental analysis. In: Market B (ed) Environmental sampling for trace analysis. Verlag Chemie, Weinheim, p 255–278
2. Dorten WS, Valenta P, Nürnberg HW (1984) Z Anal Chem 317:264–272
3. Duinker JC, Hillebrand MTJ, Nolting RF, Wellershaus S (1982) Neth J Sea Res 15(2):170–195
4. Freimann P, Schmidt D, Schomaker K (1983) Mar Chem 14:43–48
5. Guhr H, Weber E (1994) The sampling strategy in the river Elbe – experiences. In: Market B (ed) Environmental sampling for trace analysis. Verlag Chemie, Weinheim, pp 223–248
6. Haarich M, Schmidt D, Freimann P, Jacobsen A (1993) Spectrochim Acta 48B:183–192
7. Haber F (1928) Z Gesell Erdkunde, Berlin, Ergän.-Heft 3, 3–12
8. Helmers E (1991) Horizontal distribution of selected trace metals in the Atlantic ocean. Thesis, University of Bremen, pp 272
9. Helmers E, Mart L, Schulz-Baldes M, Ernst W (1990) Mar Poll Bull 21(11):515–518
10. Helmers E, Mart L, Schrems O (1991) Fresenius J Anal Chem 340, 580–584
11. Helmers E, Rutgers van der Loeff MM (1993) J Geophys Res 98(C 11), 20261–20273
12. Helmers E (1994) Fresenius Z Anal Chem 350, 62–67
13. Helmers E, Schrems O (1995) Atmos Environ 29(18):2475–2484
14. Hydes DJ, Liss PS (1976) Analyst 101:922–931
15. Klapper H, Rast W, Uhlmann D (1994) Guidlines for sampling freshwater for eutrophication management programs. In: Market B (ed) Environmental sampling for trace analysis. Verlag Chemie, Weinheim, pp 203–221

16. Koide V, Hodge VF, Yang JS, Stallard M, Goldberg EG, Calhoun J, Bertine KK (1986) Applied Geochem 1:705–714
17. Kremling K (1983) Determination of tace metals. In: Grasshoff K, Ehrhardt M, Kremling K (ed) Methods of seawater analysis. Verlag Chemie, Weinheim, pp 189–246
18. Kremling K (1985) Deep-Sea Res 32:531–555
19. Mart L (1979a) Ermittlung und Vergleich des Pegels toxischer Spurenmetalle in nordatlantischen und mediterranen Küstengewässern. Thesis, Technical University of Aachen
20. Mart L (1979b) Z Anal Chem 299:97–102
21. Mart L (1982) Talanta 29:1035–1040
22. Nolting RF, Eisma D (1988) Neth J Sea Res 22(3):219–236
23. Puls RW (1994) Groundwater sampling for metals. In: Market B (ed) Environmental sampling for trace analysis. Verlag Chemie, Weinheim, pp 287–302
24. Schaule BK (1974) Massenspektrometrische Konzentrationsbestimmung von Blei in Meerwasser. Diploma thesis, University of Heidelberg
25. Schaule B, Patterson CC (1980) The occurrence of lead in the Northeast Pacific and the effects of anthropogenic inputs. In: Branica M (ed) Lead in the marine environment. Pergamon Press, Oxford, pp 31–44
26. Schüßler U, Kremling K (1993) Deep-Sea-Res I, 40:257–266
27. Schwoerbel J (1993) Limnology. Fischer publ, Stuttgart, p 387
28. Stabel H-H, Kleiner J, Merkel P, Sinemus HW (1991) Vom Wasser 76:73–91
29. Tschöpel P, Kotz L, Schulz W, Veber M, Tölg G (1980) Fres Z Anal Chem 302:1–14
30. Whitfield M, Turner DR (1987) The role of particles in regulating the composition of seawater. In: Stumm W (ed) Aquatic surface chemistry pp 457–493

Chapter 5
Soils and Soil Solutions

Pierre del Castilho and Rainer Breder

5.1
Introduction

Soils are often extremely heterogeneous in their structure and chemical composition. A soil not only consists of solid matter of mineral or organic origin but also contains numerous pores, which altogether form a cavity system that is connected to the atmosphere. The smaller pores are filled with water, the larger ones with air and variable amounts of water. Moreover, soils contain a biotic part. As to their development and maintenance, they depend on the cooperation of bacteria, fungi, worms and insects. One function of micro-organisms is to transform plant residue into humic matter. Humic matter, but also various clay minerals and clay-humic complexes as well as manganese and iron concretions, have a high sorption capacity for potassium, ammonium, calcium, magnesium and phosphate, but most of all for heavy metals. Recent scientific information about general soil chemical aspects and the behaviour of heavy metals can be found in [1–3].

Soil pollution by heavy metals has been increasingly observed in industrialized countries. The findings led to a series of laws and decrees. Table 5.1 shows examples of threshold values for a number of potentially harmful heavy metals and arsenic in soils and soil solutions. Soil limiting values given as total metal concentrations do not provide much evidence, because there can be large differences between them and the biologically relevant concentrations. Soil solution concentrations are more relevant in this respect because dissolved heavy metals are more directly available to organisms. The actual uptake of these dissolved elements by an organism and their effects on it depend on the chemical speciation of the dissolved elements, the speed of replenishment from the solid phase to the aqueous phase, and possible synergistic and antagonistic effects. The solubility of heavy metals is influenced above all by pH changes, but also by variations of the complexing agents, the redox potential and the composition of a soil. As a rule, the element concentrations (or chemical activities) of soil solutions are governed by sorption processes which, in laboratory experiments, can be characterized by adsorption-desorption experiments [4]. Field validation of the laboratory results remains necessary for adequate predictions. Chromium and iron concentrations are not only governed by sorption reactions but also by precipitation reactions of defined compounds [4]. Copper has a high affinity for organic matter and because of this a good correlation is often found between dissolved organic carbon (DOC) and soil solution copper. Cadmium is particularly sensitive to pH changes: in only weakly polluted poor sandy soils (low in clay and organics) below pH 5.5 cadmium can already be so soluble that in some plant species the food quality standards may be exceeded [1, 5].

This chapter deals with the sampling of *total soil* and *soil solution* for subsequent trace metal analysis. The sampling of *soil air* is a particular task requiring special techniques. In this context it should only be noted that the sampling of soil air can be used for investigating soil contamination by mercury because this volatile metal has a vapour pressure high enough to enable low concentrations to be determined in soil air in a short time. Thus for this metal it is sometimes possible to restrict chemical examinations to a minimum.

In general, chemical soil investigations are chiefly performed to ascertain the following criteria:

- the influence of actual or preceding chemical inputs, waste dumps and sewage sludge disposals,
- the influences of atmospheric changes,
- the suitability of a soil with respect to the intended usage, and
- the existence of material transfers.

One condition for representative and reliable sampling is the consideration of statistical and chemometric criteria [6–10]. Their application aims at establishing the identity of the sample and the bulk material. Ultimately only an approximation as close as possible to this objective can be reached. For this purpose (among other criteria) sample selection should, for example, be performed under random conditions, meaning that preference is not given to any single part. A good approximation for the composition of the sample and the material under investigation is, of course, reached much more easily the greater the number of samples. This number also depends on the expected concentration of the elements to be determined and on their distribution. As mentioned, it is possible to look for an optimal number of samples relative to a given problem with the help of statistics and chemometrics. For further detailed information, see the relevant references [8–10].

Increments of a certain, always equal, weight should be taken from the material to be sampled. In order to calculate this weight with the prerequisite of a given statistical uncertainty and standard deviation, the distribution function of the particle sizes and also the particle shapes should be considered. The latter is practically impossible for soil samples. If in the case of an abstraction we change from squares to ellipsoids, the necessary sample mass is, for example, reduced to one half [11]. In spite of all the progress with respect to the statistical deduction and optimization of the planning of sampling, uncertainties remain which can be mimimized by the *expertise and reliability of the person performing the sampling*. In this context, examples are soil sampling with a shovel for a profile pit and also coring. These are procedures which can in particular be manually influenced and must be performed according to strictly defined rules. Whether the final product of sampling – that is the analytical sample – adequately represents the reference mass can be checked by the agreement with certain soil characteritics, for instance the grain size and the total organic carbon (TOC) content [12].

General overviews of and detailed contributions to special sampling problems can be found among the references [13–21]. An improvement of the data quality is expected in particular from the use of the international standard presently under discussion, ISO/CD 10381 Soil Quality – Sampling (Committee Draft). An international standard always represents the highest degree of standardization. ISO/CD

10381, however, still needs specific assessments for sampling guidance to be put into concrete terms.

Water in soil may be present at different depths: the unsaturated zone or the permanently saturated zone. Water from permanently water-saturated soil is called *groundwater*. The term *soil solution* is usually restricted to water from the more or less aerated topsoil with a fluctuating water content. The water content of a given soil depends on rainfall and irrigation. Groundwater – not discussed in detail here – may be studied to detect whether human activities have led to a deterioration of the quality of the raw material for drinking water. Sometimes it is investigated for the exploration of ores by means of the analysis of characteristic dissolved guide elements. Groundwater samples are collected by drilling a cylindrical hole and placing in it a pipe with a porous end deep enough to reach the relevant groundwater layer. The water is obtained by pumping at predetermined times. Locations are chosen taking into consideration the prevailing groundwater flow patterns. Currently, the design of groundwater sampling programmes, sampling techniques and the handling of water samples taken from groundwater for physical, chemical and microbiological assessment is being standardized (ISO 5667 11: 1993 Water Quality – Sampling Part 11: Guidance on Sampling of Groundwaters). More details about groundwater sampling can be found elsewhere [22].

This chapter focuses on the sampling of the soil solution in humid climatic zones (e. g. northern Europe). The study of arid zones, due to the limited amount of water,

Table 5.1. Comparison of trace metal contents in soils [1–3], groundwater and soil solutions

Trace element	Soil background values	Soil standard[a]	Groundwater standard[b]	Soil solutions (0–30 cm layer)[c] from agricultural soils	Soil solutions (0–80 cm layer)[c] from natural areas
	(mg/kg d.w.)	(mg/kg d.w.)	(µg/l)	(µg/l)	(µg/l)
Cd	0.07–0.5	1.5 (1.0)	6	0.17–3.90	0.40–27.2[c] 4.5[d]
Cr	5–100	100	30		0.5[d]
Cu	3–40	60	75	7.50–29.8	1.8 20.7[c] 60[d]
Hg	0.05–0.5	1	0.3		0.1[d]
Ni	5–50	50	75		10[d]
Pb	6–60	100	75		1[d]
Zn	10–80	200 (150)	800	6–443	10–1191[c] 5.2[d]

[a] Limiting values of the German Sewage Sludge Act of April 15, 1992, BGBL. I, p. 912. Values in brackets (Cd, Zn) for special soil conditions.
[b] Intervention values, provisional Dutch soil sanitation standards for groundwater; Circulaire Interventiewaarden Bodemsanering, published in Gazette 95 (May 24, 1994).
[c] Dutch national survey in 1991 of natural areas (woodlands, heather, grass) and soils in agricultural use; AB-DLO Haren Netherlands; to be published.
[d] Adapted from a compilation of representative natural abundances at various depths [33].

requires a different approach. Here, no direct methods to obtain the soil solution can be used, but rather indirect methods have to be used to study, for example, the mobility of heavy metals. An example is the liquid-solid extraction of heavy metals in soil using a solution of complexing agents such as EDTA or DTPA, or other standardized extraction procedures [23].

Soil solutions are analysed to (i) measure the flux of nutrients and contaminants towards ground and surface waters, (ii) estimate the phyto-available pool of nutrients and contaminants, and also to (iii) calibrate models to describe the behaviour of environmental contaminants. The soil solution is the only transport medium of non-volatile nutrients or contaminants towards plants, surface water or deeper soil layers, and groundwater has for a century been the subject of many studies, e.g. [24]. Scientists have up to now used their own strategies, sampling devices and analytical methods, thus making a comparison between results virtually impossible [25]. The application of recent insights from modern soil physico-chemistry and the standardization of methods may improve the comparability of results. Several authors have studied sewage-sludge-amended soils, e.g. [26–30]. Table 5.1 shows the heavy metal background values for soil and soil solution, and soil and groundwater standards. Also the Cd, Cu and Zn concentration ranges in the soil solution of ploughing layers of agricultural areas, and 0–80 cm layers of natural areas in the Netherlands are shown.

5.2
Materials

Generally, the sampling tools and materials as well as their treatment have to be defined by their application. If metals are used, the three least objectionable ones are titanium, aluminium and stainless steel. The best, but most expensive, material for heavy metal analysis is, of course, titanium.

5.2.1
Soil

The selection of soil sampling instruments depends on the soil type, the sampling depth and the sample mass required. A spade or shovel is used to excavate a profile pit. For drillings a lot of corer types are available. The corers can be driven into the soil by ramming, applying pressure or striking sharply. For sampling relating to trace metal determinations it is more favourable if the soil core is taken with a plastic tube inside the corer.

Box corers are also applied, allowing an undisturbed profile part to be taken for which the excavated soil cut has a defined volume which can be used, for example, for determining the density and the pore volume.

The sample containers should be made of plastic or glass and thoroughly precleaned. These containers and also the polyethylene bags should be filled so that the remaining air space is as small as possible. Moreover, the following soil sampling tools are needed: a plastic hammer and a suitable piece of wood to drive smaller box corers into the soil, a knife, a spatula, tweezers, plastic spoons and a scraping knife, de-ionized water in a syringe and a supply bottle, cellulose rags, a surveyor's rod, a topographic and a geological map, soil-science, land-use and vegetation maps, the Munsell colour charts [31], a thermometer, a camera, a portable refrigerator (12 V), a pH meter, and a balance (2–3 kg range) with battery supply.

5.2.2
Soil Solution

Soil solution sampling methods for the unsaturated zone, using field and laboratory methods, have been reviewed (e.g. Litaor, Wolt and Del Castilho [32, 33, 25]). In-situ sampling techniques make use of various of hydrophilic porous materials, each with a variety of pore sizes (distributions). The sampling probes are placed in the soil for equilibration for about six months to one year before sample collection. The probes may sample gravitational water only (sampling of freely flowing water) or after a filling period at reduced tension they may sample by suction a mixture of free water and the more firmly held water in the fine soil capillary pores. The probes may have the shape of a cup, tube (including hollow fibres) or plate, and are made of a variety of materials: nylon, PVC, porcelain, etc. The most serious problems with field sampling devices are (i) maintaining uniform permeability of the probes in space and time, and (ii) the differences in effectiveness of sampling under saturated vs. unsaturated soil conditions. Both affect the spatial and physical origin of the water collected. Because of the large variety of probe types and materials, standardization of this type of sampling seems difficult.

Taking soil core samples in the field and obtaining the soil solution in the laboratory is a more promising way towards standardization. The most frequently used methods are the preparation of a saturated soil paste with the subsequent separation of the liquid phase, and the extraction of soil samples by centrifugation.

The saturated paste method [34] has been standardized for the purpose of measuring the soluble salt content of soils. A soil sample is wetted up to the saturation point and after equilibration the water is obtained by suction. The original soil salt content (chloride, nitrate) can be calculated using the relevant dilution factor. The method is less suitable for cation concentrations that depend on the ion strength (heavy metals). The method may, however, be modified by using a salt solution (with salt concentrations typical of the particular field) to prepare the saturated paste instead of distilled water.

The centrifugation methods presently in use are based on the work of Edmunds and Bath [35] and have recently been recommended by Bufflap and Allen in a comparison of four pore water sampling techniques (squeezing, vacuum filtration, dialysis, centrifugation) [36]. These authors studied the effect of centrifugal force and duration, and the effect of soil pore size on the soil solution collection efficiency. A variety of centrifugal forces are used in work reported in the literature. At present there is a tendency towards the lower rotational speeds because at such speeds the composition of the soil solutions seems less method-dependent than at high speeds. Centrifugal displacement for 15–30 minutes at low pressures (< 500 kPa) is now often used [33].

The centrifugation method seems to be standardizable as only a few parameters need proper description: soil sample dimensions, type of rotor (swing out, fixed angle), centrifugal force and time, and soil-water interface (type of filter paper). Once these parameters have been described, the reproducibility and repeatability of the method can be fixed. The materials used for the construction of the centrifuge assembly may depend on the type of analyte. For heavy metal research, plastic, aluminium, stainless steel and titanium are preferred. Losses of dissolved heavy metals by sorption of the filter paper may be expected, especially in samples with a soil solution of pH > 5. In such cases the filter paper should be in contact with the field-moist soil sample some time before soil centrifugation.

5.3
Aspects of Soil Sampling and Recommendations for Its Realization

At first it should be noted again that mistakes made during sampling are, of course, not correctable. The steps treated here are shown in Table 5.2.

During soil sampling, both for soil and soil solution examination, the strategy used is important. In particular this concerns two questions: how many samples should be taken and where to obtain representative results. Unfortunately in most cases, economic reasons often are a limiting factor for the sampling strategy. With geostatistical procedures such as variogramme analysis and kriging the number of sampling points within a surveyor's grid may be reducible without any loss of essential information about the background load [36, 37]. Kriging means the attempt to obtain further relevant data by appropriate smoothing and interference procedures. In this case, the data of the surrounding places are quasi-estimated [38] for those places for which no results are available.

In the chronological sequence of sampling, first a plan has to be set up. Then no working step may be taken on an accidental basis. Whereas for naturally grown soils a relatively homogeneous distribution of trace metals can be considered, for anthropogenically polluted soils (e.g. urban or industrial regions) a very heterogeneous distribution must be expected.

To become acquainted with the history of the location, registers of real estate or other archives should be studied and the occupants can be interviewed. Careful documentation is a major objective in all attempts to regulate and standardize sampling. The complete statements used in the sampling report are presented in Sect. 5.3.1.

Table 5.2. Sampling steps

Step	Examples for consideration and research
1 Planning	Definition of the objectives, sample number and size, method selection
2 Preparation of sampling	Compilation of logistical and technical tools, cleaning of containers and instruments
3 Sampling	Random or systematic selection of sampling sites, boreholes or trial pits, characterization of the site and the sample
4 Sample storage	Storage time, place and temperature

The *soil type* can be established according to the naturally formed horizons, which can be recognized in a borehole 0.5–2 m deep. The soil type has to be determined in any case. For instance, the distribution of substances that are characteristic of various soil types, such as degraded loss soil (Lessivé) and podzol, should be considered. Also the choice of sampling tools has to be harmonized with the soil types. With respect to podzol, for example, the formation of firm boundary stones (iron pan) in the illuvial horizon is possible by the shift of iron compounds. If pollutants have accumulated on such layers, then a borehole has to be well covered for safety. Otherwise the pollutants may be rinsed into the deeper soil and contaminate the groundwater. If considered necessary, a cover (e.g. of bentonite) can be placed around the break point. The sampling of a gley in the groundwater region needs special support devices for the corer that prevent "leaking" of the core. Because the excavation of a profile pit causes quite a lot interference in a soil, it should be done with care and installed only after pre-tests at particularly obvious places have been made. Excavation should only exceed 1.2 m in exceptional cases because otherwise, for security reasons, a slope would have to be constructed, which requires a great deal of space with increasing depth.

When a profile pit is filled in again, material suspected of pollution should be covered again with unpolluted soil. If a repetition of sampling is planned, the place should be permanently marked and a site plan made.

The sampling depth for arable land usually reaches down to the bottom of the ploughing layer (30 cm). With respect to meadows, sampling is only performed down to a depth of 10 cm because of the plain root distribution. The depth of interest for children's playgrounds is 0 to 0.35 m, and for sports grounds and recreational areas 0 to 0.15 m. If intervals are necessary, they may be equal or may differ between these sampling depths, depending on the special objective and the local conditions. Metric sampling may also be appropriate in cases where the soil profile is not greatly differentiated [21]. The investigation of contaminated sites, where information is needed from deeper layers, may even require a sampling depth of several metres.

The number and distance of *lateral samples* depends on the size of the area to be examined and e.g. the extent and degree of pollution. If a soil has been the substratum for different plants or has been treated with various fertilizers, or the area consists of different soil types, then an average sample should be taken from each of the sections, because of the different geological parent material.

In general the sampling points can be arranged at one's own discretion, if, for example, a polluted area is directly discerned, randomly or in systematic patterns (Fig. 5.1). The spacing of the pattern is chosen according to the desired information content or the required security of the statement or the financial budget.

At the sampler's discretion the sample choice is marked by subjective criteria that are not reproducible. The quality of an expert-based pattern depends above all on the knowledge the person in charge has of the sampling site. The random grid, however, is not largely influenced by subjective criteria, but of course not reproducible either.

With respect to simple systematic patterns, the sampling points are arranged at equal distances. In comparison to square grids, triangular grids are advantageous (Fig. 5.2) because the size of the unsampled circular area decreases perceptibly. Also the probability of detecting directed structures is essentially enhanced.

Judgement

Random

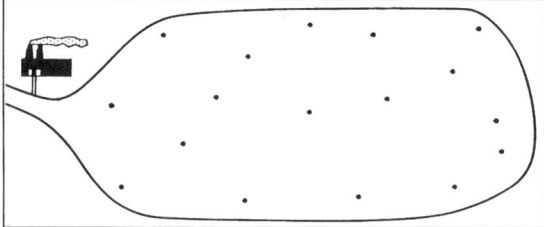

Systematic

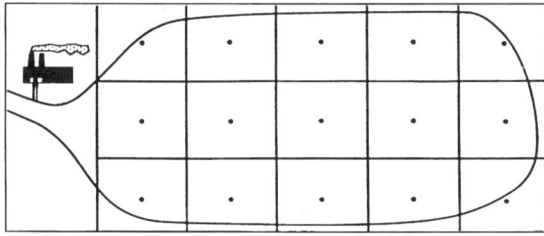

Fig. 5.1. Possible arrangement of sampling points (after [18])

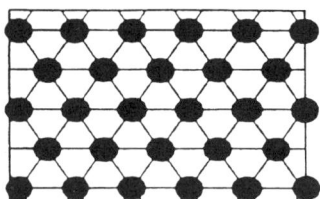

Fig. 5.2. Triangular grid

If soil examinations along line emitters are necessary, the potential pollution gradients must first be known in order to find the optimal distance of the sampling points from the emission source. This pattern is used to detect the soil pollution along actual or potential line emitters such as roads, railway lines, pipelines, supply lines, etc.

If pollution should be unearthed, or in the case of diffuse pollution, sampling can be achieved with fewer points by placing them in an X- or W-shape (Fig. 5.3).

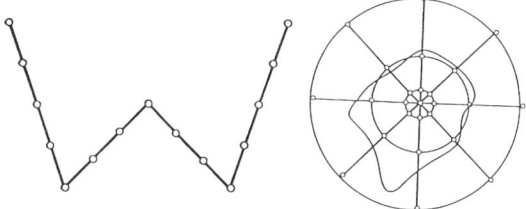

Fig. 5.3. Patterns for the survey of suspect areas (W-shape and circular grid)

Sampling along two diagonals (X-shape), however, unfavourably weights the sample towards the middle of the area.

In the case of a point source and unknown spreading of pollution, a distribution of sampling points on concentric circles is indicated (Fig. 5.3). With respect to the whole area, more sampling points are located at the centre of such a pattern.

The production of *composite samples* is used to attain interpretable results in spite of the high small-dimensional variability of soil properties. If composite samples are taken, it must be ensured that the utilization and soil relationships are fairly homogeneous within the area represented by the composite sample. The profile structure, the substrate, the content of organic carbon etc. should not differ significantly within the sampled area. The number of increments for one composite sample varies from 15 to 50 and may represent 1 ha, 5 composite samples 15–20 ha of a fairly homogeneous site [21]. The increments must be taken according to volume or mass equivalent and can be pre-homogenized in plastic tubes. Samples of different horizons should not be mixed.

The *sample mass* is dependent on the nature and number of the intended measurements (subsamples) as well as on the size of the soil particles and should be sufficient for homogenization. The perpetual problem for the analyst is not to fall short of the minimum sample mass. For the determination of heavy metals and accompanying parameters, 1 kg of fresh soil is sufficient in the case of fine-grained clay-rich material. Sandy soils and extraction procedures may need a larger amount.

To avoid *contamination* it is possible to drive the core into a plastic tube or pipe inside the sampling device. If no plastic insert is used, the core should be peeled with a plastic scraper. That is also relevant for trial pits dug with a spade.

Cross-contamination from sample to sample is avoided if after every sampling the walls of the instruments are cleaned of adhering soil remnants, rinsed with distilled water and dried with tearproof pulp sheets. In general the sampling tools and containers must not separate the elements to be determined.

5.3.1
Soil Sampling Report

The sampling report (minutes) should comprise the following information:

- sample number;
- name of the sampler;

- date of sampling;
- weather, past weather (in particular date of the last rainfall);
- characteristics of the sampling site (co-ordinates, inclination, layout plan, actual exploitation – different exploitation at the sampling site should be considered according to the size);
- sampling tools;
- profile pit or drilling hole (a profile or trial pit should only be excavated down to a depth of 1.2 m, otherwise expensive safety measures such as escarpments are necessary);
- inner diameter of the corer;
- sampling depth above/below the mineral soil surface;
- soil horizon (upper, lower limit);
- soil type (single sample or composite sample with the number of sampling points and site area);
- mass and, if applicable, volume of the sample (diameter and height of the box corer)
- homogeneity of the sample;
- occurrence of alien elements;
- colour (assessment with Munsell colour charts);
- smell;
- sample transport (vessel material, temperature, duration of transport);
- sample storage (duration, place and temperature).

5.3.2
Details of Soil Solution Sampling

Bulked or non-bulked soil samples may be centrifuged, or water extracted, to obtain a solution for chemical analysis. Figure 5.4 shows a typical centrifuge tube assembly for soil centrifugation. For non-bulked soil solution sampling, undisturbed soil cylinders are placed directly in the centrifuge tube and tared before centrifugation. For bulked sampling of the soil solution, field-moist soil is placed in the tube in small portions and subsequently pressed down with a piece of wood or plastic. After filling the tubes, they are weighed and tared before centrifugation. During centrifugation the soil solution is expelled and is collected in a vessel at the bottom of the device. For metal analysis an appropriate amount of acid is added to the collection vessel immediately after centrifugation (pH 1–2).

The minimum number of samples required depends on the variation in the results that can be tolerated. The results may be biased by the method and time of sampling. The latter is affected by fluctuations in the recent history of the soils: rainfall and treatments such as fertilization or manuring. For instance, both soil water content and agricultural practices affect the salt content of the soil solution, and consequently the concentration of those analytes which depend on ion strength. The heavy-metal soil solution concentrations of a particular soil depend on a combination of solid-phase parameters and the concentration of dissolved ligands from, for example, dissolved organic matter, the concentration and type of dissolved salts, pH and redox potential [3]. Therefore, for monitoring the quality of the soil solution taken from agricultural soils and natural areas, sampling should

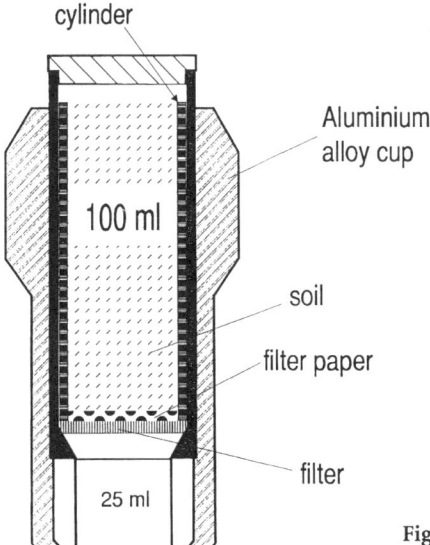

Fig. 5.4. Typical centrifuge tube assembly for soil centrifugation

always be repeated at the same time of the year, and before manuring/fertilization. To further improve the reproducibility of the results, standardization of sampling methods and further procedures should be encouraged.

In humid climatic zones, there is particular interest in the composition of the soil solution in the winter-spring period, at a depth beneath the rooting zone. At that time of the year, the dissolved nutrients and contaminants move down towards the groundwater, and, depending on the soil physico-chemical factors may not return to the topsoil layers. Knowledge of concentrations, soil chemical processes, and hydraulic parameters is essential for soil element balance calculations.

5.4
Sample Storage

Soil samples must be cooled directly after being taken. Among other things, cooling and freezing reduce redox and enzyme reactions, and the activity of microorganisms. The temperature of cooling or freezing depends on the intended time of storage and the degree of unchanged preservation of the sample desired.

5.4.1
Soils

At room temperature soils can be stored only in the dry state, which is best achieved by freeze-drying. At ambient temperature air-dried soil contains about 3–5% water, freeze-dried soil less than 3%. The freeze-dried samples are gently rolled to break up soil aggregates and sieved through plastic sieves with round holes of 2 mm diameter to remove e.g. stones, roots and twigs. The samples are ground in agate mills and stored in precleaned glass bottles in a cool, dark and dry place. If after several months or years, metal species such as tributyltin or methylmercury have to be determined, it is necessary to store the original samples in the gas phase over liquid nitrogen, as performed in environmental specimen banking [39], see also Chapters 2 and 7 of this book.

5.4.2
Soil Solutions

Once the soil solution has been obtained by centrifugation or another method, it can be analysed as such, or after membrane-filtration (e.g. 0.45 or 0.20 µm) or super-centrifugation (e.g. 40,000 g, 1 hour) to remove suspended materials, depending on the purpose of the investigation. Sample conservation is important as many analytes tend to sorb onto the walls of containers, filter paper, etc. Sorption losses may become a problem, especially when the soil solution and the soil are no longer in contact. Losses of analytes can be minimized by common methods used in aquatic science (see Chapters 1, 3 and 4 of this book). For heavy metal determination samples are usually acidified to pH 1–2.

Soil solution samples or extracts should preferably be analysed immediately after collection, but they can also be stored after collection and pre-treatment. The material of the storage containers depends on the type of analytes involved. For heavy metals, polypropylene or other plastic materials free from heavy metals are often used in combination with sample acidification to pH 1–2 and storage in the dark at 4 °C.

5.5
Quality Control

Here only some general remarks will be made because the quality control and quality assurance systems that already exist for the analytical laboratory are not yet available for soil sampling. References to quality control can, of course, be found in other chapters of this book, among others particularly concerning to the comprehensibility of documentation as performed in entire and detailed protocols. At the present time, on the basis of the ISO 9000–9004 or EN 45001 – EN 45003, EN 45011 and EN 45012, sampling guidelines and strategies can be developed. They have to be adapted to the specific sampling problem. If this is done, the results of soil surveys can also be used and interpreted by other workers, even years later.

In the following, the sampling steps for metal determinations in soils and soil solutions particularly relevant for quality control are summarized:

- selection of the optimum structure of the sampling grid;
- choice of appropriate equipment;
- site, profile and sample description;
- production of composite samples (number and mass of the single samples),

- preservation and transport;
- storage without changes, e.g. contamination or losses.

It should be mentioned that with fresh soil samples or soil solutions to which for example, known amounts of a certain metal were added, transport, storage and also matrix effects could be tested, analogous to the study of reliability of the analytical method with certified reference materials.

5.6
Safety Precautions

An extensive survey on safety precautions is presented in "Committee Draft: Soil Quality – Sampling – Part 3: Guidance on Safety – ISO/DIS 10381-3".

Persons performing sampling operations must also be acquainted with the relevant national safety regulations. In particular this includes a preliminary investigation of the sampling site in order to locate actual or potential dangers, e.g. type of contamination, flammable gases, energy supply lines, covered hollow spaces, etc. Also knowledge of the hazards arising from the handling of equipment and chemicals is necessary. During sampling, gloves and dust protection or breathing masks should be worn. If there are pieces of broken glass or nails, safety shoes are preferable to boots. A first aid kit must be available. Special precautions should be taken during the sampling of soils on waste disposal sites (e.g. release of methane that can form explosive gas mixtures or release of the poisonous hydrogen sulphide). In the presence of residual pollution from armaments, the brave sampler should take special care with his drilling apparatus, otherwise he might be a late victim of former political disasters.

5.7
References

1. Kabata-Pendias A, Pendias H (1992) Trace Elements in Soils and Plants. CRC Press Boca Raton, Florida, USA
2. Scheffer F, Schachtschabel P (1992) Lehrbuch der Bodenkunde. Enke, Stuttgart
3. Salomons W, Stigliani WM (1995) Biogeodynamics of Pollutants in Soils and Sediments. Springer, Berlin Heidelberg New York
4. Welp G, Zheng Y, Brümmer GW, Breder R (1991) Mitt Dt Bodenkundl Gesellsch 66:427
5. Del Castilho P, Chardon WJ (1995) Plant and Soil 171:263
6. Minkkinen P (1987) Anal Chim Acta 196:237
7. Lamé FPJ, Defize PR (1993) Environ Sci Technol 27:2035
8. Gy PM (1992) Sampling of Heterogeneous and Dynamic Material Systems. Elsevier, Amsterdam
9. Webster R (1985) Quantitative Spatial Analysis of Soil in the Field. In: Stewart BA (ed) Advances in Soil Science, Vol 3, pp 1–70, Springer, New York Berlin
10. Kateman G (1987) Chemometrics – Sampling Strategies. In: Topics in Current Chemistry Springer, Berlin Heidelberg New York
11. Brands G (1983) Fresenius Z Anal Chem 314:6
12. Schwartz V (1995) Fresenius J Anal Chem 351:629
13. Kratochvil B, Wallace D, Taylor JK (1984) Anal Chem 56:113R
14. Gomez A, Leschber R, Hermite PL (1986) Sampling Problems for the Chemical Analysis of Sludge, Soils and Plants. Elsevier Applied Science Publishers, London New York

15. Berrow ML (1988) Anal Proc 25:116
16. Melcher RG, Peters TL, Emmel HW (1986) Sampling and Sample Preparation of Environmental Material. In: Topics in Current Chemistry. Springer, Berlin Heidelberg New York
17. Fortunati GU, Banfi C, Pasturenzi M (1994) Fresenius J Anal Chem 348:86
18. Keith LH (1990) Environ Sci Technol 24:610
19. Rubio R, Ure AM (1993) Intern J Environ Anal Chem 51:205
20. Fränzle O (1994) Representative Soil Sampling. In: Markert B (ed) Environmental Sampling for Trace Analysis, pp 305–320, VCH, Weinheim
21. Paetz A, Crößmann G (1994) Problems and Results in the Development of International Standards for Sampling and Pretreatment of Soils. In: Markert B (ed) Environmental Sampling for Trace Analysis, pp 321–334, VCH, Weinheim
22. Nash RG, Leslie AR (1991) Groundwater Residue Sampling Design. ASC Symp Series 465 Am Chem Soc, Washington, DC
23. Quevauviller Ph, Lachica M, Barahona E, Rauret G, Ure A, Gomez A, Muntau H (1996) Sci Total Environ 178:127
24. Briggs LJ, McCall AG (1904) Science 20:566
25. Del Castilho P (1994) Quim Anal 13 [Suppl 1]:S21
26. Behel D, Nelson DW, Sommers LE (1983) J Environ Qual 12:181
27. Emmerich WE, Lund LJ, Page AL, Chang AC (1982) J Environ Qual 11:182
28. Gerritse RG, van Driel W, Smilde KW, van Luit B (1983) Plant and Soil 75:393
29. Hodgson JF, Geering HR, Norvell WA (1965) Soil Sci Soc Am Proc 29:665
30. Sidle RC, Kardos LT (1977) J Environ Qual 6:431
31. Munsell Soil Color Charts, Munsell Color Company, Baltimore, USA
32. Litaor MI (1988) Water Resour Res 24:727
33. Wolt JD (1994) Soil Solution Chemistry: Applications to Environmental Science and Agriculture. Wiley, New York
34. Rhoades JD (1982) Soluble Salts. In: Pape AL et al. (eds). Methods of Soil Analysis, Part 2. Chemical and Microbiological Properties. Agronomy Monograph no 9 (end ed) ASA-SSA, Madison, WI, USA, pp 167
35. Edmunds WM, Bath AH (1976) Environ Sci Technol 10:467
36. Bufflap SE, Allen HE (1995) Wat Res 29:2051
37. Einax J, Machelett B, Geiß S, Danzer K (1992) Fresenius J Anal Chem 342:267
38. Tao S (1995) Water, Air and Soil Pollution 83:161
39. Haas CT (1992) Krig User's Manual, University of Wisconsin, Milwaukee
40. Stoeppler M, Zeisler R (ed) (1993) Biological Environmental Specimen Banking. Special Issue, Sci Total Environ 139, 140

Chapter 6
Waste

ULRICH OSBERGHAUS and ECKARD HELMERS

6.1
Introduction

Waste may be defined as "moveable property which the possessor wishes to get rid of or which has to be disposed of in an orderly manner to maintain the public good, particularly environmental protection" [1]. *Waste* is a comprehensive term. It includes e.g. industrial waste, domestic refuse, rubbish, excavated material, waste oil, wrecked cars, used tyres, used galvanic solutions, sewage sludge and numerous other materials.

Waste can be solid and liquid with numerous shades in between. Domestic refuse is an especially heterogeneous solid material.

Waste can be transported in different conditions and packages either continuously in waste streams (conveyor belts etc.) or discontinuously in transport vehicles.

Certain types of waste may be reworked into new products, or incinerated as a fuel. They may be defined as "residual matter that requires monitoring" (German Refuse Act, Sect. 2 Para. 3).

Waste is examined in order

- to establish its suitability for recycling,
- to allocate the waste to certain utilisation procedures and utilisation plants,
- to define an appropriate disposal site or treatment plant,
- to determine the calorific value,
- to obtain fundamental knowledge about waste type and composition.

In order to obtain information to answer these questions, sampling and analysis of waste is required at different positions within the cycles of use of a material (Fig. 6.1).

The examination of sewage sludge further aims at:

- testing its suitability for agricultural application,
- identifying persistent chemical compounds from indirect discharge (i.e., discharge into a sewer),
- monitoring the success of municipal and industrial efforts to decrease the burden of waste water,
- testing the operational status of waste water purifying plants.

The most important prerequisite for an examination of waste in general and of sewage sludge in particular is professional sampling. Analytical results obtained by expending a great deal of time and money are useless if it is not known with any certainty whether the sample can be considered as representative of the parent population.

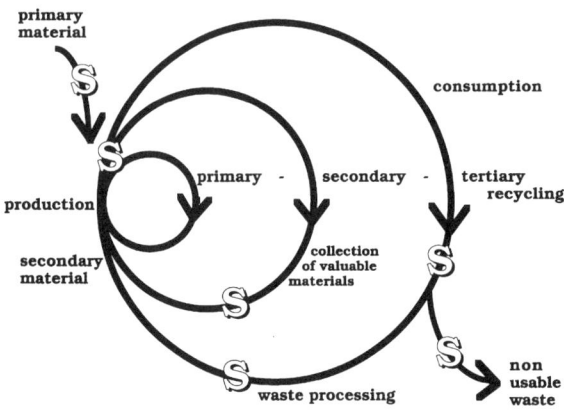

Fig. 6.1. Location and need for analytical sampling ("S") in materials and cycles of use, according to [18]

Since raw waste material possesses the highest grade of heterogeneity, the chemical composition is preferably monitored in the products of waste processing. As a consequence, information on the raw material is usually focused on physical properties (e.g. paper, glass, metals) while most of the available analytical data deal with the processing products (liquid or solid residues), as for example incineration ashes or exhausts [2, 3] (see Sect. 6.6).

6.2
Theoretical Considerations for Sampling

6.2.1
General Terms

The terms to be used in connection with sample collection and sample preparation have not yet been standardised. They are selected in accordance with Kraft [4], Park and Pohland [5] and Garfield [6].

From the *parent population, samples* are collected which constitute the *bulk sample*. Ideally, equal portions must be taken at random points, throughout the entire batch, with a sufficient number n of points sampled. As the amount of the bulk sample will mostly be larger than is required for analysis, it has to be *reduced* by repeated grinding, sieving and subdivision (e.g. quartering) to yield the *laboratory sample*, or *sample for analysis* (reduced sample; test portion; see Fig. 6.2).

It is necessary that

- the bulk sample represents the parent population,
- the laboratory sample for analysis represents the bulk sample.

6.2.2
Deduction of a Criterion for Representativeness

It is assumed that the analytical value of the parent population under consideration is normally distributed. Thus, the mean μ and the standard deviation σ of the

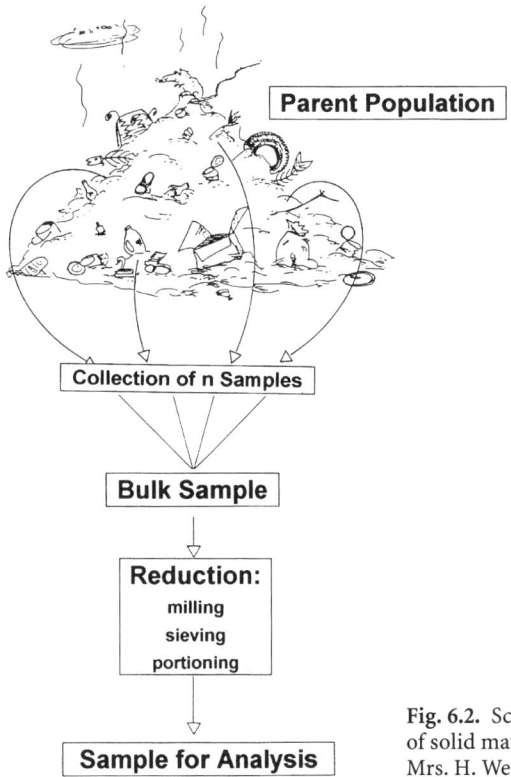

Fig. 6.2. Scheme for sample collection of solid material. (With kind permission of Mrs. H. Weyers, Aachen, Germany)

analytical value can be attributed to the parent population. On a limited number of n samples – the bulk sample – the same number of analyses are performed. The result is a frequency distribution with the characteristic criteria mean $\bar{x}$ and standard deviation s of the analytical results x of n samples.

It is expected that the bulk sample taken from the parent population represents this parent population for the property of interest. The representativeness of a sample may be quantified by the degree and the reliability of approximation of the obtained mean value $\bar{x}$ as compared to the true mean μ of the parent population (see e.g. [7]).

The confidence interval $(\bar{x} - d) \leq \mu \leq (\bar{x} + d)$ to be derived from Student's t distribution with

$$d = \frac{t_{n-1;P} \cdot s}{\sqrt{n}} \tag{6.1}$$

includes the mean μ of the parent population on the basis of the selected statistical certainty P (see e.g. [8]). For P values of 90%, 95% or 99% are generally assumed. The factor $t_{n-1;P}$, whose values are tabulated, becomes smaller with an increasing number of samples, n, and with decreasing statistical certainty P; the confidence limit changes similarly.

The confidence interval $2d$ may thus be considered as an inverse measure of representativeness: representativeness will decrease with increasing values of the confidence interval, and vice versa.

Representativeness as specified here may refer to an area (or a volume) as well as a period of time. *Reference to a location* means that a number n of samples is collected at one point of time on n sampling points; the resulting bulk sample will represent the geometrical entity which may be, for example, a truck load or a refuse dump. *Reference to a period of time* means that n samples will be collected subsequently at one well-defined sampling point.

It has to be mentioned that the demand for representative sampling of solid waste (e.g. municipal waste) may pose severe or even insuperable problems. A common problem lies in the existence of extreme values within a limited series of data. These extreme values may be due to "hot spots" in a waste sample, or to an exceptional event (e.g. discharge into the sewer, as in the case of sewage sludge) within the considered period of time. Special care has to be taken over the evaluation and presentation of a set of data containing extreme values (see Sect. 6.2.6). Also long-term trends may overlay the normal scattering of a data set (see Sect. 6.6.2).

6.2.3
Relation Between Sampling Error and Analytical Error

The scatter of the analytical value x depends on the errors of sampling and that of the analytical method, given as relative standard deviations $s_P/\bar{x}_P$ and $s_A/\bar{x}_A$ or (each multiplied by 100) as variation coefficients. Both result in the total deviation of the analytical value $s/\bar{x}$ according to

$$(s/\bar{x})^2 = (s_P/\bar{x}_P)^2 + (s_A/\bar{x}_A)^2 \tag{6.2}$$

(see e.g. [9]). In the case of solid samples, particularly solid waste, the sampling error usually exceeds the analytical error significantly.

6.2.4
Variables which Affect the Sampling Error

After Wilson [10], the relative sampling errors $s_P/\bar{x}$ for a binary mixture of two granulated compounds 1 and 2 with a mean content $\bar{x}$ of component 1 in the mixture can be estimated as

$$\frac{s_P}{\bar{x}} = \sqrt{\frac{(1-\bar{x})}{\bar{x}} \frac{\rho_1 \cdot \rho_2}{\bar{\rho}^2} \frac{\bar{V}}{V_P}} \tag{6.3}$$

with $\bar{V}$ mean volume of a grain,
      V_P volume of the sample,
      ρ_1, ρ_2 densities of components 1 and 2,
      $\bar{\rho}$ mean density of the basic material,
      s_P standard deviation of the contents x in n samples.

Waste

Following the assumption that the densities of components 1 and 2 differ only insignificantly, then after the introduction of the mean mass of a grain, $\bar{m}$, and the total mass of the sample, m_P, it follows for the relative sampling error that:

$$\frac{s_P}{\bar{x}} = \sqrt{\frac{(1-\bar{x})}{\bar{x}} \cdot \frac{\bar{m}}{m_P}} \qquad (6.4)$$

From this it is obvious that the sampling error increases

- with decreasing sample mass m_p or decreasing sample volume V_p,
- with increasing mean grain mass $\bar{m}$ or increasing mean grain volume $\bar{V}$.

With $V \approx a^3$, where a is the diameter of the grain, it follows from Eqs. (6.3) and (6.4) after raising to the second power:

$$\left(\frac{s_P}{\bar{x}}\right)^2 \approx \sqrt{\frac{(1-\bar{x})a^3}{\bar{x} m_P}} \qquad (6.5)$$

or

$$m_P \left(\frac{s_P}{\bar{x}}\right)^2 \approx \sqrt{\frac{(1-\bar{x})a^3}{\bar{x}}} \qquad (6.6)$$

The functions $s_P/\bar{x} = f(a)$ for m_P = const., or $m_P = f(a)$ for $s_P/\bar{x}$ = const. result in a straight line in the double logarithmic coordinate system. Under practical conditions one can observe families of parallel straight lines (Figs. 6.3 and 6.4).

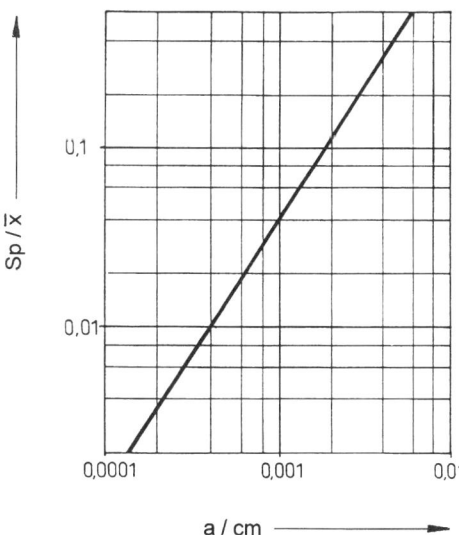

Fig. 6.3. Sampling error as depending on the grain size for the example of tin ore, according to Baule and Benedetti-Pichler. Modified from [7]. Sample mass 5 mg. (With kind permission of "Deutscher Verlag für Grundstoffindustrie", Leipzig)

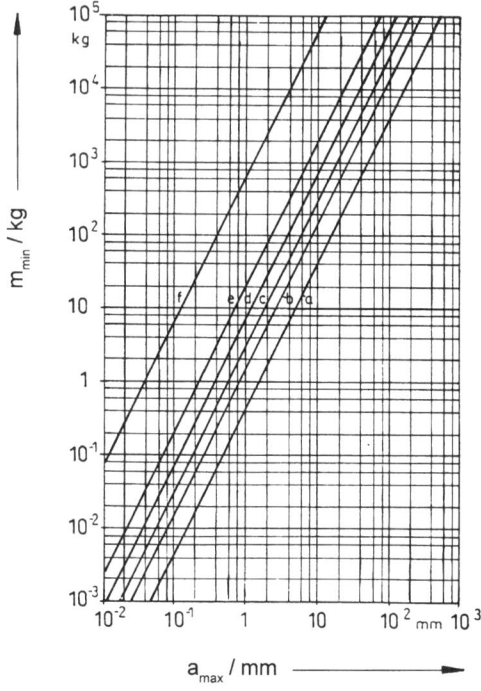

Fig. 6.4. Minimum mass of the bulk sample as depending on the grain size of the largest grain, according to Taggart (from [4]); a–f, homogeneity parameters

Consequently for a given grain size distribution of the material to be examined, the analyst has two options, either

- to define the weight of the bulk sample (e.g. 1 kg) and thus to accept that the sampling error can reach any unknown magnitude, or
- to accept a certain sampling error, and to adapt the sample amount accordingly.

6.2.5
Estimation of the Required Number of Samples

The value of the confidence interval, which includes the true mean value μ of a (normally distributed) analytical value, depends after Eq. (6.1) on

- the selected number of samples, n,
- the selected statistical certainty P, which decisively determines the factor $t_{n-1,P}$, as well as,
- the experimentally determined scatter of the analytical value x, given by the standard deviation s that is, as already mentioned, composed of sampling error and analytical error.

Rearrangement of Eq. (6.1) gives

$$n = \left(\frac{t_{n-1;P} \cdot s}{d}\right)^2 \tag{6.7}$$

Waste

The required number of samples, n, thus follows from the smallest tolerable representativeness, i.e. the largest tolerable confidence interval $(\bar{x} - d) \leq \mu \leq (\bar{x} + d)$.

6.2.6
Examples

Some of the terms used so far will be elucidated using the example of sewage sludge as a comparatively homogeneous waste.

In the framework of an extensive investigation on behalf of the German Environmental Specimen Bank in Jülich, FRG, the prerequisites and requirements for representative sampling of sewage sludge were to be established on 9 sewage treatment plants of different sizes [11, 12]. Samples were collected at one definite point each in monthly or bimonthly intervals in the course of almost 3 years and analysed for 7 heavy metals and for 15 polycyclic aromatic hydrocarbons.

The results will be interpreted using the example of the lead content in sewage sludge from one plant that processes the collected wastewaters of a large German city located on the river Rhine. Three concentration-time-series for lead are presented in Fig. 6.5.

The results for plant STA as derived from 11 individual data were as follows:

average value $\bar{x} = 1820$ mg/kg, standard deviation $s = 357$ mg/kg

Representativeness: As a sewage treatment plant represents a steady-state system with a continuous output of sewage sludge, the term *representativeness of the bulk sample* refers to the period of time monitored, and not to an area or a volume (see Sect. 6.2.2).

With $\bar{x} = 1820$ mg/kg and $s = 357$ mg/kg, the half confidence interval d was computed by means of Eq. (6.1) as

$d = 240$ mg/kg,

using the value of $t = 2.228$, tabulated for a confidence level of $P = 95\%$ (double-sided approach) with 10 degrees of freedom. The resulting upper and lower confidence limits, x_U and x_l, have been entered into Fig. 6.5.

The result may be interpreted for the example of lead as follows:

The confidence interval 1580 to 2060 mg/kg contains, on a confidence level of 95%, the true average lead concentration of the sewage sludge discharged within the period March 1986 to February 1987.

Required number of samples: On the other hand, a confidence interval may be set in correspondence to the desired representativeness, yielding a certain number of samples, n, on the basis of an empirically found standard deviation. Let us again consider the lead content of the sewage sludge discharged from the treatment plant under concern. The standard deviation was found to be $s = 357$ mg/kg, corresponding to a coefficient of variation of 19.6%. Let the requirement be a higher representativeness as reflected by a half confidence limit of $d = 182$ mg/kg corresponding

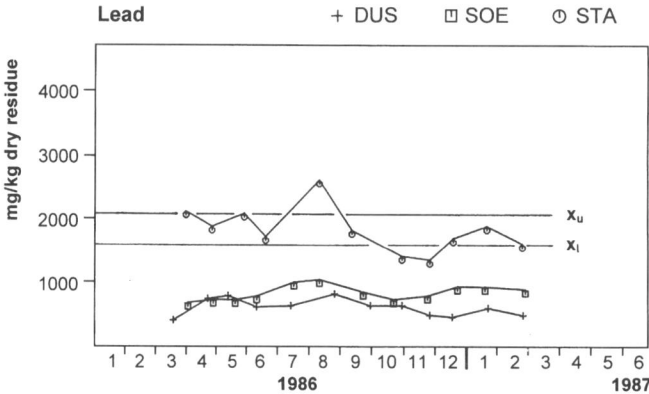

Fig. 6.5. Concentration of lead in sewage sludges from 3 different large-scale sewage treatment plants over one year [11]

to 10% of the average value. The number of samples follows from Eq. (6.7) with $t = 2.228$ as $n = 19$.

The result may be interpreted as follows: *19 samples are required in order to enclose the true average lead concentration of the sewage sludge discharged in the period March 1986 to February 1987 within the confidence interval of 1638 to 2002 mg/kg.*

Obviously, this is a requirement which one will find difficult to meet in practice. Selection of a clearly larger confidence interval of, say, ± 25% (i.e., $d = 455$ mg/kg) around the estimated average value yields a required number of samples of $n = 3$. No more than 3 samples have to be collected within the period to be represented, provided the less strict requirements are accepted.

The approach so far elaborated also serves as a basis for surface water sampling programmes [13, 14] and may be used to establish sample collection programmes on a reasonable and comprehensible basis. In the present context, it applies to comparatively homogeneous waste discharged from steady state-systems as may be found in an industrial process or a sewage treatment plant. The standard deviation to be derived from the test period should be based on a sufficient number of samples.

One precondition for the applicability of Eqs. (6.1) and (6.7) is that the parent population – i.e. the composition of the individual samples – be normally distributed. As far as sewage sludge is concerned, this precondition does not always hold. In some cases, individual extreme concentration values or highly structured concentration courses may be found [11, 12]. Take the example of plant SET as presented in Fig. 6.6. An extreme value exceeding 20 mg/kg for mercury was found in September 1986. Calculation of the confidence interval or the required sample number will not yield reasonable results. The causes for the observed effects should rather be analysed, and an adapted sampling programme be derived, depending on what the data are supposed to reflect. If such a data series has to be summarized, it may be more sensible to present the median value than the average value, as the

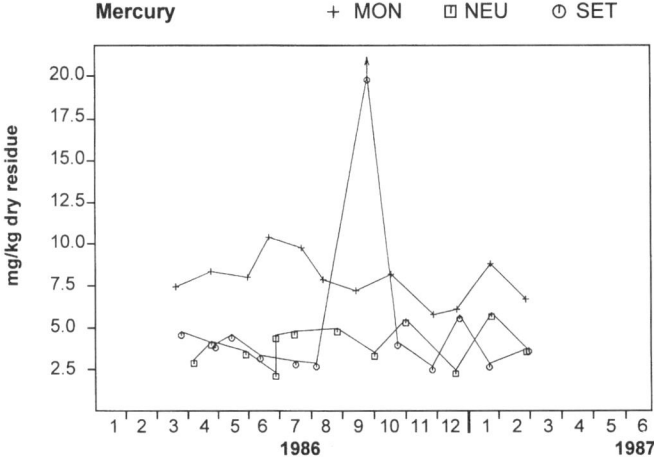

Fig. 6.6. Concentration of mercury in sewage sludges from 3 different medium-scale sewage treatment plants over one year [11]

median value is far less affected by extreme values than the average value. Alternatively, one may calculate the average value, the confidence interval or the number of samples required on the basis of the homogeneous (outlier-free) set of data, and present the extreme value as individual result.

As a further precondition for the applicability of Eqs. (6.1) and (6.7), the compositions of subsequent samples should not be *autocorrelated*; i.e. they have to be independent. Take a sewage treatment plant as an example of autocorrelation: as the average residence time of sewage sludge in a digester is of the order of 2 weeks, the compositions of digested sludges collected on subsequent days will be mutually linked or autocorrelated. In this case, a suitable sampling interval will be one month or more.

6.3
Sampling and Storage

The tools and containers for sample collection are as varied as the types of waste to be collected [15, 16]. Solid and loose waste close to the surface can be collected in the simplest case with a shovel or a spatula. If the waste occurs in larger layers or at greater depths, it is necessary to use different types of sample corers and borers or even heavy tools such as dredging machines. Also Pürckhauer borers, split spoon samplers and other tools familiar to geological investigations are in use.

Scooping instruments are used for the sampling of liquids and fluid sludges at shallow depths. Scoops fixed on rods are the most suitable and easy to handle. Pipettes are cylindrical sampling tools of various lengths that allow sampling along a profile of selected depth. Sludge excavators, immersion bombs and immersion cylinders simply fixed on drag ropes can be used to collect liquid samples from any depth.

Solid sludges and pasty materials can be collected with sampling tubes. Screw type, groove-shaped or sleeve sampling tubes have to be employed in such cases depending on the purpose.

Further information on sampling tools can be found in a recent booklet published by the British Royal Society of Chemistry [17], the German LAGA guideline PN2/78 K [16] and, in connection with standard specifications, in [15].

Closable wide-neck vessels of sufficient volume are common as sample containers. The material of these containers and of the sampling tools should not contaminate the sample with the compounds to be subsequently determined, nor should they absorb these compounds. They should not corrode during sampling and sample storage.

The manifold problems of wall effects will be addressed only briefly here. If for example plastic bottles are used, the inner surface may absorb significant amounts of oil and grease or chlorinated hydrocarbons during sample storage. Just the reverse may happen if an oil sample is stored in a plastic bottle containing softeners which can diffuse into the oily phase. Glass vessels are well suited for most materials, however, they do not allow freezing of liquid samples, e.g. sludge. Vessels made of metal, e.g. of aluminium, allow freezing of the sample and safe sample transport, but are not resistant to corrosive liquids.

6.4
Decomposition and Analysis

The decomposition method to be applied for elemental analysis depends on the analytical parameter, the matrix, the extent of anthropogenic contamination and the legal requirements laid down (e.g. in a number of EPA instructions that may be found in [19]).

Depending on the type of physico-chemical binding between the matrix and the elements to be analysed, simple leaching, pressure decomposition with acids or dry ashing in a muffle furnace may be required. Nitric acid, aqua regia or hydrofluoric acid may be necessary if pressure decomposition is to be employed. Additionally, oxidising agents such as hydrogen peroxide or perchloric acid may be recommended.

Some general indications on the decomposition of waste will be given in accordance with [19]. Since alkali and alkaline earth elements are less volatile, they may be analysed after dry ashing of the sample. In order to enable elemental analysis of sewage sludge or soil, samples may be dissolved in aqua regia and boiled under reflux. To ensure the availability of silicate-bound elements during analysis, the samples must be treated with hydrofluoric acid followed by dissolution into hydrochloric and nitric acids.

A – possibly sequential – elution may be performed with demineralised water, acidified water at selected pH steps, or with artificially mineralised water [19].

As the composition of waste is often highly heterogeneous, the amount of sample taken for analysis becomes a crucial frame condition. In general, if there is a choice, a decomposition device suitable for larger sample amounts is preferred.

Because of the high concentration levels that usually need to be analysed, the subsequent instrumental analysis of elemental concentrations may be performed with standard atomic spectroscopy equipment. In most cases, the sensitivity of the techniques based on optical emission spectrometry (ICP-OES) or flame atomic absorption spectrometry (F-AAS) will be sufficient. Hydride-forming elements (e.g. As) and Hg can be detected by appropriate hydride AAS techniques.

Further elements (e.g. in incineration products) can be measured by coulometry (C, S), photometry (P) or ion-selective electrodes (F).

6.5
Legal Requirements, Standards and Instruction Leaflets

Regulations and rules concerning waste are rare and often arbitrary and contradictory. As standardized sampling procedures exist for various industrial bulk goods including foodstuffs, it is common practice to adapt these procedures to the sample collection of waste.

The German Refuse Act (Sect. 4 Para. 5 AbfG) [1] is the basis of the Technical Instructions on Waste (TA Abfall) whose first part is now legally in force [20]. In Appendix D, the Technical Instructions lay down 25 limiting values for waste to be disposed of in landfill sites. The limiting values refer to various physicochemical and chemical properties of waste samples as may be the loss on ignition or the cadmium content in waste leachate. For sampling procedures, Appendix B refers to two guidelines drawn up by the "Waste" working group of the German Federal States (LAGA) [16]:

- PN2/78K: basic rules for sampling of waste and dumped materials (as of Dec. 1983);
- PN2/78: collection and preparation of samples of solid, sludgy and liquid waste (as of May 1979).

The guidelines introduce several sampling rules which are relatively easy to realise. They refer to solid and liquid waste, solid sludge, residues from the incineration of domestic refuse and other matter, and are intended for practical work by decision makers on-site.

However, the guidelines quoted are lacking any scientific basis, yielding an unpredictable sampling error. For example, the minimum mass of a waste sample to be collected is fixed at 0.06-fold (in kilogrammes) of the maximum grain size (in mm). A maximum grain size of, say, 30 mm thus yields a minimum sample mass of 1.8 kg. As Rasemann [18] points out, this rule has been adapted, without any justification on statistical grounds, from the German standardized procedure DIN 51701 concerning hard coal of grain sizes > 120 mm.

Further indications regarding the sampling procedure in general as well as the required sample amount in particular may be adopted from standardized procedures which have been established in the context of industrial products and raw materials, for instance iron ores (ISO 3081), solid mineral fuels (ISO/TC 27), agricultural foods (ISO/TC 34), fertilizers (ISO/TC 134), or ceramic material (DIN 51061; all standards quoted according to [18]).

Because of its heterogeneity and grain size distribution, municipal waste occupies a special position. Therefore, a considerable effort is necessary to obtain representative samples. Jäger [21] estimates that a representative bulk sample should comprise approximately 1 % of the parent population. A valid investigation of the composition of domestic refuse should be based on a sampling period of one week. An average weekly sample amount of 7500 kg obtained from 1500 inhabitants, will represent approximately 150,000 inhabitants.

Instruction leaflet M4 "Refuse Analyses" from the Association of Municipal Transport Fleets and Cleaning Services (VKF) and the "Study Group for Municipal Waste Management" (AKA) [22] envisages a refuse vehicle which passes through several districts of the town according to a fixed sampling plan and collects a set number of refuse containers from each district in order to obtain an average refuse sample. The size of the refuse sample collected in this way should be "at least 1000 kg".

Examination of the unsorted bulk sample of municipal waste is of interest in a few cases only. Examples are the determination of parameters such as dry residue, loss on ignition or calorific value. In most cases sampling will be followed by sieving which separates the bulk sample into different grain-size fractions. The instruction leaflet M4 proposes fractionation into

- fine refuse   (< 8 mm)
- medium refuse   (8 – 40 mm)
- coarse refuse   (40 – 120 mm) and the
- residue from sieving (> 120 mm).

The coarse fraction, which is defined in different ways by different authors [16, 21, 23], is subsequently sorted by hand into the classes of paper, glass, plastics, metals, plant material, etc. Fractions from sieving and from sorting are prepared and analysed separately.

The German Sewage Sludge Directive (Klärschlammverordnung) [24] regulates the application of sewage sludge to agriculturally or horticulturally used soils (Sect. 1, Para. 1, AbfKlärV). According to Appendix 1 of this directive, the collection of sewage sludge samples should be performed in accordance with DIN 38414-S1 [25]. The problem of obtaining representative analytical results is also addressed in some detail: as a supplement to DIN 38414-S1, Appendix 1 of AbfKlärV prescribes that five litres of sludge must be collected "from each of at least five different sewage sludge batches and combined in suitable containers (e.g. made of aluminium) for a composite sample. Sample collection should, if possible, be performed on different days". A subsample will then be collected from the composite sample and forwarded to the laboratory.

Since these legislative and practical examples are based on German laws and directives, some remarks should be made on the legislative basis for the handling and disposal of hazardous waste within the European Union. According to the CEC (EU) Directive on Toxic and Dangerous Waste [26] the member states of the EU must ensure that toxic and hazardous waste is kept separate from other matter, that the packaging is appropriate and that the labelling indicates its composition and origin. At the disposal site, the composition of hazardous waste must be recorded. A listed number of toxic and hazardous substance requires priority consideration, such as antimony, arsenic, beryllium, cadmium, chromium (VI), lead, mercury, selenium, tellurium, thallium and their compounds, metal carbonyls and soluble copper compounds. Within the framework of this directive, detailed regulations for the management of hazardous waste have been enforced in a number of EU member states. Similar regulations have also been issued in many countries outside the EU. Detailed information about these regulations and also concerning the export of waste to other countries outside the European Union can be found in [19] and [27].

6.6
Specific Problems

6.6.1
Monitoring Metal Concentrations in Municipal Waste and Incineration Residues

The chemical analysis of municipal waste is a challenge not only due to the variety of existing materials. Concentrations of pollutants in municipal waste (Fig. 6.7) fluctuate over several orders of magnitude.

Additionally, the distribution and location of "hot spots" may be extremely irregular within a waste segment [19].

According to different physico-chemical properties, such as volatility, selected indicator elements (C, P, S, F, Cl, Fe, Zn, Pb, Cd, Cu, Hg) are usually analysed [2]. The behaviour of Fe, for example, is representative of the lithophilic elements (including e.g. Co, Mn, Cr). Cd, on the other hand, possesses the characteristic properties of an element of higher volatility (also Sb, Sn, and Th). Hg, with one of the highest volatilities, should be treated separately.

The products of the incineration process are slag, filter ash, filter cake, wastewater and exhaust gas. Their composition is more uniform than the composition of municipal waste. Consequently, wherever waste treatment has to be monitored, it is mostly the treatment products which are analysed in order to trace the transport and distribution of toxic elements during the treatment process.

The transfer of five metals through a municipal solid waste incinerator is shown in Table 6.1. Though the metal portions found in the different incineration products will vary with the specific technique under consideration, the different volatilities of the elements are well reflected.

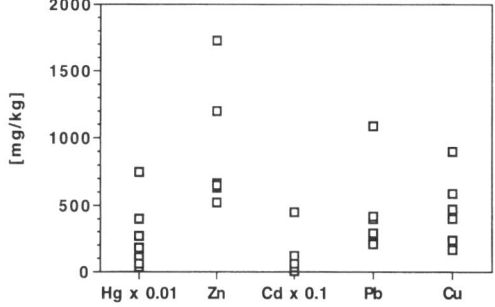

Fig. 6.7. Concentrations of five elements in solid municipal wastes; means each from eight different investigations, data from [2, 32, 33]

Table 6.1. Transfer of five metals through a municipal solid waste incinerator (concentration in solid municipal waste = 100 %) to different incineration products (%). Data from [3 and 32]

Metal	Incineration products				
	Slag	Filter ash	Filter cake	Waste water	Exhaust gas
Cu	94	6	< 1	< 1	< 1
Pb	72	28	< 1	< 1	< 1
Cd	9	91	< 1	< 1	< 1
Zn	46	54	< 1	< 1	< 1
Hg	10	20			70

The element composition of municipal waste may be indirectly derived from the composition of the incineration products [2]. By taking into account

- the mass balance of the incineration products under consideration, and
- the degree of scattering of concentration values of the elements under concern in the considered incineration products,

one may calculate

- an optimum sampling frequency as well as
- the uncertainty (i.e. confidence interval) of the element concentration in the municipal waste.

For 9 mass and trace elements, Table 6.2 presents the optimal ratio of the number of analyses required for slag and filter ash over an investigation period of 72 h [2].

The ratios represent the specific incineration process under consideration. They reflect, on the one hand, the different physicochemical properties (e.g., volatilities) of selected elements, leading to typical distributions in the two incineration products. They reflect, on the other hand, the mass balance for slag and filter ash in the incineration plant. They finally reflect the degree of scattering of concentration values for the 9 mass and trace elements being investigated.

The resulting 95 % confidence intervals for various element concentrations (F, P, S, Fe, Cu, Zn, Cd, Hg, Pb) in municipal waste from Vienna were found to range typically between 7 and 20 %. This is surprisingly low compared to the uncertainty obtained when bulk samples of municiple waste are analysed after repeated grinding, milling, sieving and quartering.

Table 6.2. Optimal ratio of the number of analyses required for slag and filter ash over an investigation period of 72 h, calculated for a specific incineration plant [2]

Element	F	P	S	Fe	Cu	Zn	Cd	Hg	Pb
Ratio	9	22	5	38	61	2	0.1	0.4	7

6.6.2
Elemental Analysis of Sewage Sludge and Sewage Sludge Ash

In contrast to municipal waste, sewage sludge constitutes a relatively homogeneous matrix with a considerably lower fluctuation of pollutant concentrations. Due to the long detention time of the sludges in the treatment plant, one single grab sample is related to the input into a sewage treatment plant over nearly two weeks [28]. Consequently, it is much easier here to sample in a representative way. On the other hand, the basic assumption of random fluctuations of elemental concentrations does not necessarily hold (see 6.2.6) [11, 12]: Firstly, input into the sewerage system is strongly correlated with meteorological conditions. For example, large amounts of pollutants may be transported into the sewer system during heavy rainfall or within a short period of time after an extended frost period. Secondly, the anthropogenic input may fluctuate, largely as a consequence of changes in factory production (holidays, closures). Moreover, there may be long-term variations (trends) in the anthropogenic input, for example due to the phasing-out of leaded petrol (Fig. 6.8, taken from [26]).

In addition to Pb, there has been a long-term decrease in the elements Ag, Cd, Cr, Cu, Ni, Sn and Zn. On the other hand, there has been an increase of Sb, Mo, Ti, Sr, Pt and Rh in sewage sludge, as has been shown, for example, at the main sewage treatment plant of Stuttgart [29 and unpublished results of E. Helmers].

Since the concentrations of selected metals in sewage sludge have exceeded the legal limits regulating the agricultural use during the past few decades, incineration has become one of the most popular removal techniques. Prior to the dumping of the resulting ashes, elemental concentrations are routinely controlled. Because of the enrichment of most elements in the ashes, the fluctuation of elemental parameters in sewage sludge can be monitored in a very sensitive way [29].

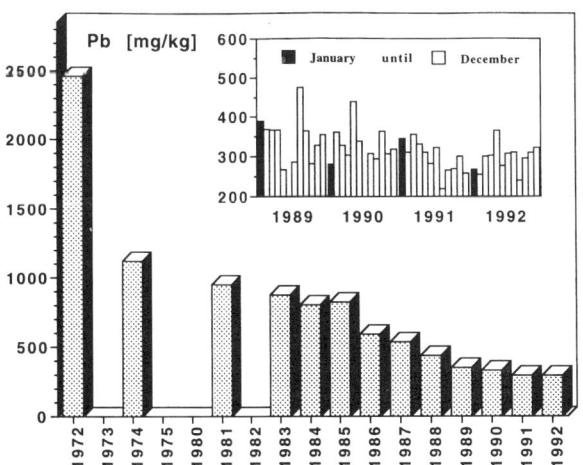

Fig. 6.8. Variations of lead concentrations (means) in sewage sludge ash from year to year (1984 to 1992) and from month to month (1989 to 1992, small picture)

Table 6.3. Metal and metalloid species in gases from sewage sludges and domestic waste deposits [30, 31]. Me = methyl, Et = ethyl, Bu = butyl

As	Bi	Hg	Pb	Sb	Sn	Te
AsH_3	$BiMe_3$	$HgMe_2$	$PbMe_4$	SbH_3[b]	SnH_4[b]	$TeMe_2$
$AsMeH_2$			$PbEtMe_3$[a]	$SbMe_3$	$SnMe_2H_2$	
$AsMe_2H$			$PbEt_2Me_2$[a]	$SbEt_3$[b]	$SnMeH_3$	
$AsMe_3$			$PbEt_3Me$[a]		$SnMe_4$	
$AsEtMe_2$			$PbEt_4$		$Sn(n\text{-}Bu)H_3$	
$AsEt_3$					$Sn(n\text{-}Bu)_2H_2$	

[a] Detected in sewage gas only, [b] Detected in landfill gas only.

6.6.3
Metal and Metalloid Species in Gases from Sewage Sludges and Domestic Waste Deposits

Municipal waste may not only be incinerated but also disposed of in landfill. Microbiotic activity changes the composition of these deposits significantly. The process which is accompanied by the release of large quantities of gas commonly reaches a steady state after two years [19]. As a further consequence of microbiotic activity, organometallic species are formed which can be detected at trace levels in gases from sewage sludges and domestic waste deposits (Table 6.3). The investigation of these possibly toxic species is an example of advanced inorganic speciation analysis.

The metal species are collected at –80 °C by a cryogenic glass trap which is subsequently mounted into a gas chromatograph. The analysis is then performed by real-time on-line coupling of a low-temperature GC with ICP-MS. Both the chromatograms and the mass spectra are recorded in order to identify the single species by their retention times, boiling points and masses. Semi-quantification is achieved by interelement calibration, yielding concentrations between 0.003 (Se) and 71.6 (Sb) µg/m³ [30, 31].

6.7
References

1. Gesetz über die Vermeidung und Entsorgung von Abfällen (Abfallgesetz – AbfG) vom 27. August 1986 (BGBl. I S. 1410), zuletzt geändert durch das Einigungsvertragsgesetz vom 18. September 1990 (BGBl. II S. 885, 1990)
2. Bauer G, Brunner PH (1996) Müll und Abfall 1:19
3. Bauer G, Schachermayer E (1996) Environ Sci Poll Res 3:10
4. Kraft G (1980) Probennahme an festen Stoffen. In: Kienitz H, Bock R, Fresenius W, Huber W, Tölg G (eds) Analytiker Taschenbuch. Springer, Berlin Heidelberg New York
5. Park DL, Pohland AE (1989) J Assoc Off Anal Chem 72:399
6. Garfield FM (1989) J Assoc Off Anal Chem 72:405
7. Doerffel K (1984) Statistik in der analytischen Chemie. Verlag Chemie, Weinheim
8. Kreyszig E (1979) Statistische Methoden und ihre Anwendungen. Vandenhoek und Ruprecht, Göttingen

9. Doerffel K, Eckschlager K (1981) Optimale Strategien in der Analytik. Verlag Harri Deutsch, Thun Frankfurt a. M.
10. Wilson D (1964) Analyst 89:18
11. Osberghaus U, Stoeppler M, Grimmer G (1992) Beiträge zur Umweltprobenbank – Nr. 8: Untersuchung des Konzentrations-Zeit-Profils ausgewählter Stoffe in Klärschlämmen – Ableitung einer Strategie zur Probenahme von Klärschlämmen definierter Repräsentanz. Berichte des Forschungszentrums Jülich, Jül-2691; ISSN 0366-0885
12. Osberghaus U, Stoeppler M, Grimmer G (1994) Beiträge zur Umweltprobenbank – Nr. 9: Ermittlung und Interpretation von Konzentrationstrends für ausgewählte Summenparameter und Spurenkomponenten in Klärschlämmen hoher Repräsentanz. Berichte des Forschungszentrums Jülich, Jül-2976; ISSN 0944-2952
13. Craenenbroek Van W, Smits A, Baadof B, Meijers A (1985) Das Gas und Wasserfach 126:486–492
14. ISO 5667/1 (1981) Wasserbeschaffenheit – Probennahme – Teil 1: Richtlinien zur Aufstellung von Probennahmeprogrammen. In: Vom Wasser 56:288–312
15. DIN 51750 (1990, 1991) Prüfung von Mineralölen/Probennahme, Teil 1, Allgemeines, Dezember 1990. Teil 2, Flüssige Stoffe, Dezember 1990. Teil 3 Salbenartig-konsistente und feste Stoffe, Februar 1991. Normenausschuß Materialprüfung im Deutschen Institut für Normung e. V.
16. Hösel G, Schenkel W, Schnurer H (1984) Müll-Handbuch, Lieferung 2/1984, Kennziffer 1859/1860/1861, Erich Schmidt, Berlin
17. Crosby NT and Patel I (1995) General principles of good sampling practice. The Royal Society of Chemistry, Cambridge, UK
18. Rasemann W (1993) Müll und Abfall 6:460–469
19. Rump HH, Scholz B (eds) (1995) Untersuchung von Abfällen, Reststoffen und Altlasten. Verlag Chemie, Weinheim
20. Gesamtfassung der Zweiten allgemeinen Verwaltungsvorschrift zum Abfallgesetz (TA Abfall). Teil 1: Technische Anleitung zur Lagerung, chemisch/physikalischen, biologischen Behandlung, Verbrennung und Ablagerung von besonders überwachungsbedürftigen Abfällen vom 12. März 1991 (GMBl S. 139, 1991)
21. Jäger B (1988) Bestimmung der Zusammensetzung fester Abfälle. In: Straub H, Hösel G, Schenkel W (eds) Müll- und Abfallbeseitigung, Lieferung 3/1988, Kennziffer 1710/1720. Erich Schmidt, Berlin
22. Hösel G, Schenkel W, Schnurer H (1988) Müll-Handbuch, Lieferung 3/1988, Kennziffer 1710/1729. Erich Schmidt, Berlin
23. Gorbauch H, Rump HH, Schneider W (1986) In: Spillmann P (ed) Wasser und Stoffhaushalt in Abfalldeponien und deren Wirkungen auf Gewässer. Verlag Chemie, Weinheim
24. Klärschlammverordnung (AbfKlärV) vom 15. April 1992 (BGBl. I S. 912, 1992)
25. DIN 3814-S1 (Nov. 1986) Deutsche Einheitsverfahren zur Wasser-Abwasser- und Schlammuntersuchung/Schlamm und Sedimente/Probennahme von Schlämmen. Normenausschuß Wasserwesen im Deutschen Institut für Normung e. V.
26. Council Directive on Toxic and Dangerous Waste (78/319/EEC). OJ No L 84 of 1.03.1978
27. Ewers U (1991) Standards, Guidelines, and Legislative Regulations Concerning Metals and Their Compounds. In: Merian E (ed) Metals and Their Compounds in the Environment. VCH Publisher, Weinheim, pp 687–711
28. Helmers E, Wilk G & Wippler K (1995) Chemosphere 30:89–101
29. Helmers E, Mergel N & Barchet (1994) UWSF – Z Umweltchem Ökotox 6:130–134
30. Feldmann J, Grümping R, Hirner AV (1994) Fresenius J Anal Chem 350:228–234
31. Feldmann J, Hirner AV (1995) Intern J. Environ Anal Chem 60:339–359
32. Heintz A, Reinhard G, Chemie und Umwelt. 2nd Ed, Vieweg Publisher, Braunschweig 1991
33. Förster U, Colombi C, Kistler R (1991) Dumping of wastes. In: Merian E (ed) Metals and Their Compounds in the Environment, VCH Publisher, Weinheim, pp 333–355

Chapter 7
Collection, Preparation and Long-Term Storage of Marine Samples

JOHANN-DIEDERICH SCHLADOT and FRIEDRICH BACKHAUS

7.1
Introduction

The continuous monitoring of pollutants in water, soil and air, as well as on different trophic levels of the food chain (plants, animals), which leads at last to man himself, is one of the main goals of environmental protection policy. However, these inquiries are not restricted to monitoring stationary pollutions. In a special sense they require information about the behaviour of these substances in the environment. The following questions arise:

- Where do these substances remain and where do they accumulate?
- Which short- or long-term effects do they have on man and the environment?
- When and where do these substances first enter the environment?
- Which new substances do the environmental chemicals change into and over what time span?
- Do potentially new toxic substances originate and how stable are these substances?
- In which chemical form do they occur?
- How mobile are these substances in the environment?

The accumulation of pollutants in terrestrial and floral tissue allows changes in the burden to be recorded by selecting suitable specimens. These specimens consequently have an indicator function in the sense of early recognition of hazardous chemicals in the environment (see Fig. 7.1).

Using the knowledge and methods available today, only insufficient predictions are possible concerning the influences and effects of artificially produced and intentionally or unintentionally released hazardous chemicals on pristine ecosystems. Likewise, predictions and estimations of the potential risks for the health and well-being of mankind are possible [1].

According to the "*European Inventory of Existing Commercial Substances*" *EINECS* (GDCh/BUA) [2] at present about 100 000 different chemical substances are produced year in year out. Their behaviour and the effects within the environment are still largely unknown [3]. The major tasks are:

- to detect even very small changes within the environment by special investigations and *monitoring*,
- to elucidate the reasons for these changes and,
- if need be, to counteract their progress.

At the same time attention is focussed on the protection of man and the environment against anthropogenic and geogenic pollutants and their systematic and con-

Collection, Preparation and Long-Term Storage of Marine Samples

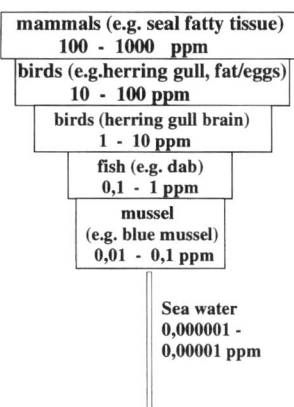

Fig. 7.1. Accumulation of polychlorinated biphenyles (PCBs) in organisms of the marine environment

tinuous recording in soil, water, and air, as well as in selected biological samples from the foodweb.

7.2
Tasks of the Environmental Specimen Bank

When characterizing and evaluating environmental and also human samples in their *present state* – and their development, too – important assumptions have to be made:

- to recognize current deleterious developments,
- to estimate the manner and extent of existing deleterious developments and their consequences,
- to produce evidence for the implementation of political measures, and
- to work out a basis for the policy of the Federal Government in the conservation of nature and the environment, as well as for the health and well-being of mankind.

These objectives led to analytical studies in 1976. Between 1979 and 1984 the logistics and the technical requirements for the realization of an Environmental Specimen Bank were worked out within a pilot project, "Umweltprobenbank", financed by the German Federal Ministry of Research and Technology (BMFT) [4].

7.2.1
Institutions Involved

Since January 1985 the ESB has been a permanent programme within a division of the Federal Ministry of the Interior (BMI) (later on within the *Federal Ministry of the Environment, Nature Conservation and Nuclear Safety, BMU*). The name *Um-*

weltprobenbank des Bundes (Federal Environmental Specimen Bank) combines two specimen banks:

- *Probenbank für Umweltproben* (Specimen Bank for Environmental Samples), located in the Research Center of Jülich in the Institute of Applied Physical Chemistry, and
- *Probenbank für Humanorganproben* (Specimen Bank for Human Tissue Samples), located at the University of Münster.

Additional institutions support the tasks of the Environmental Specimen Bank:

- *Lehrstuhl für Biogeographie* (Chair of Biogeography) at the University of the Saarland,
- *Institut für ökologische Chemie* (Institute of Ecological Chemistry), GSF Neuherberg,
- *Biochemisches Institut für Umweltcarcinogene* (Biochemical Institute of Environmental Carcinogens), Großhansdorf.

7.2.2
Selected Representative Ecosystems in Germany

Operating the "Umweltprobenbank des Bundes" [5, 6, 7] is part of the "ecological environmental monitoring" of biological samples in their ecosystematic context, while is still under construction in addition to the already existing air and water monitoring programmes.

An expert commission of the Ministry for the Environment, Nature Conservation and Nuclear Safety selected different ecosystem types for a continuous concept. Since 1985 collections have been made every two years in these areas (see Fig. 7.2). It is planned to change the two-year sampling frequency into a yearly sampling frequency by the year 2000; by doing this, the analytical characterisation data can be used more effectively for monitoring purposes. Since German re-unification in 1989, materials have been collected more intensively in the "new federal states". Collection work is carried out with the support of local institutions.

A representative selection of matrices from the terrestrial, limnic and marine environments, as well as from human beings, is collected, as follows:

- soil, tree samples (leaves and shoots), animal samples (pigeon eggs, liver of roe deer, and earthworms) from the *terrestrial* environment;
- sediment, fish (muscle and liver) and zebra mussels from the *limnic* environment;
- sediment, algae, common mussels, fish (muscle and liver) and eggs of sea birds (see Fig. 7.3) from the *marine* environment.
- Additionally, human samples are collected and analysed for both organic as well as for inorganic substances.

The use of the materials for retrospective analysis is only possible because they are deep frozen to cryogenic temperatures directly at the collection site. Possible chemical changes are avoided or at least minimized [8, 9].

Special guidelines (*Standard Operating Procedures, SOP*) have been developed for the collection, as well as for the further preparational steps [10].

Collection, Preparation and Long-Term Storage of Marine Samples

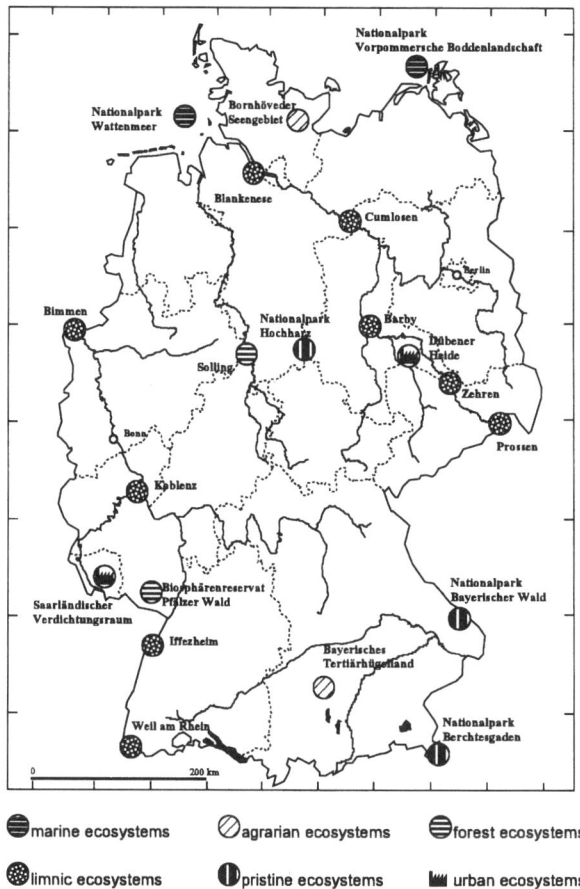

Fig. 7.2. Selected representative areas of the Federal German Environmental Specimen Bank

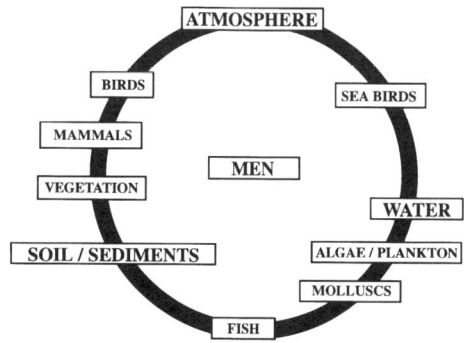

Fig. 7.3. Selected specimens from different environmental compartments

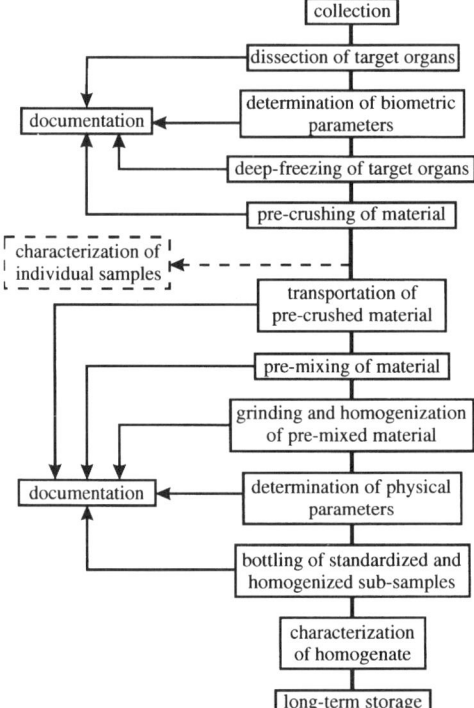

Fig. 7.4. Sample pretreatment for environmental samples of the ESB

The preparation of the material (processing, homogenizing and bottling of standardized subsamples) (see Fig. 7.4) is carried out under permanent control in the above-mentioned temperature range.

7.3
Collection

7.3.1
Collection Principles

Collection is the first and most important step in chemical, physical and biological investigations and is therefore of highest significance. Errors made during the collection phase cannot be corrected even by the very costly measures described in the following process steps.

The analyst has to rely on contamination-free, well-documented sampling by the collection team. In the past, analysts themselves carried out all the steps from collection to analysis and evaluation or at least they checked all these steps. Today this is nearly impossible because of specialisation within the analysis procedures. Besides, samples are generally analysed in different laboratories. This is applicable to all ESB activities because it is not certain which substances will be analysed in future. In the first characterization of ESB samples only essential and non-essential elements, organo-metal compounds, chlorinated hydrocarbons (CHCs) and polycyclic aromatic hydrocarbons (PAHs) are investigated.

Therefore possible contamination by trace elements as well as by organic substances has to be avoided or minimized. This will exclude the use of certain plastics and metal tools. Depending on the matrices, PTFE tools or stainless-steel tools and in some special cases titanium and quartz glass tools are used.

In order to lower the detection limits very expensive analytical tools often have to be purchased. However, it is very difficult to justify such financial expenses for technical improvements in collection and sample preparation. The awareness of the need for financial support for sampling and sample preparation has to be raised.

In order to ensure reliable collection work, stringent requirements have to be met:

- deployment of experienced personnel;
- regular training of the collection personnel;
- provision of appropriate technical tools, according to the state of the art;
- development and supply of detailed written guidelines, according to the latest operating procedures. For the German ESB, appropriate guidelines (*Standard Operating Procedures* "*SOP*") are published for different matrices, for the preparational steps and storage as well as for analytical characterization [11];
- development of sample protocols that identify all the characteristics of an area; e.g. weather conditions, collection data and possibly differences from the SOP.
- regular communication between collection personnel and analysts, according to the HACCP concept (*HACCP* = *H*azardous *A*nalysis *C*ritical *C*ontrol *P*oint).

The last requirement is of particular importance and therefore in the following further details of the HACCP concept are given:

- information for the collection personnel about the method to be used for spatial and temporal representativeness;
- identification of potential contamination sources;
- working out the sampling design in co-operation with collection and analytical personnel;
- the collection team has to prepare a protocol for each collection which describes the implementation of the collection in as much detail as possible so that there are no doubts about the evaluation of the analytical values;
- if necessary the collection has to be elucidated with the aid of photographs and sketches.

The setting up a new collection area should preceded by a search of the literature with the aim of discovering whether there are other research activities with similar interests, or if the planned activities represent additional investigations. In general, all investigations in a new area should be performed according to a well co-ordinated concept for all research groups. This will minimize the "scientific disturbance" in those selected areas.

Criteria for *"temporal representativeness"* are:

- time of the collection;
- duration of the collection;
- frequency of the collection.

The *"spatial representativeness"* of a collection is characterized by the following parameters:

- site of collection (assignment of priorities after implementation of special screenings in the selected areas);
- location of collection sites in the selected areas.

The *"sampling techniques"* parameters are of special importance:

- type of collection (documentation);
- type and material of tools used;
- cleaning of tools and containers used (instructions documented in special guidelines).

Sampling has to be carried out according to the latest guidelines. Possible differences from the status of the sampling guidelines have to be documented in the sampling protocols in great detail.

Intensive training of the collection team is the basic requirement for a representative collection: *The collection team has to be aware that their task is a very important part of the whole investigation!*

7.3.2
Collection of Marine Matrices

Possible contamination has to be considered depending on the substances investigated. Investigations of ultra-trace metals in water require special care with respect to contamination by personnel and tools [12]. Special pre-cleaned sample containers in double- or triple wrapping should only be opened using gloves at the site and should be secured for transport in the same way in a triple wrapping [13, 14]. The gaseous volume above the sample itself – "dead volume" – should be as small as possible.

Collection of *sediment samples* should not involve a separation of the very fine fraction (< 20 µm) directly by wet sieving. In the case of sediment sampling the whole fraction should be collected and homogenized. The fractionation should be done directly before the analytical clean-up procedure. Sediment samples should be deep-frozen directly at the collection site to avoid chemical changes caused by the reaction of suspended matter with air. The freezing of glass bottles with sediment samples requires a special freezing technique: the glass bottles are frozen while being continuously rotated in liquid nitrogen at an inclination of 60°. This will avoid cracking of the glass during the deep-freezing of the sample.

The collection of macro-algae – bladder wrack algae (Fucus vesiculosus) – can be done periodically at an interval of two months in order to investigate seasonal variations at one location and to be able to collect the new tips of the fronds (thalli). The sampling of the macro-algae is carried out at the collection site by taking randomly selected plants representatively over the whole site.

The young fronds should be cut off using Teflon®-coated pair of scissors. Afterwards the macro-algae are cleansed of any adhering animals or other plants, if necessary. To avoid chemical changes the macro-algae may be deep frozen directly.

For other investigation purposes the samples may be stored in PE bags or PE bottles. In the laboratories the macro-algae are prepared further.

Common mussels (Mytilus edulis) should be collected at least twice a year in late autumn and spring from natural mussel beds. Mussel beds on the North Sea coast are dependent on tidal influences. Usually mussels of these mussel beds are collected manually at low tide. At least 70 individuals should be collected and homogenized to fulfil statistical prerequisites [15]. About 250 mussels are usually collected from the whole mussel bed area and combined to one homogenate. The population of the mussel bed is assumed to be individuals per m^2. The size of the mussels collected should be 4–7 cm according to the collection guidelines [16] and the Trilateral Monitoring Expert Group (TMEG) agreement of the common Wadden Zee secretary [17].

Natural mussel beds in the Baltic Sea are always underwater (–5 to –15 m), so that the mussels are generally collected using a dredge, which is made of stainless steel and nylon net to minimize contamination. The amount of mussels collected by dredges is about 10–15 kg.

Mussels should be cleaned with water from the collection site. Afterwards the water should be allowed to drain off for several hours (at least 2). In some cases, for special investigation purposes, a defaecation process in clean sea water is necessary. For transportation the mussels should be frozen; to avoid chemical changes, deep-freezing is suitable. The following biometric parameters are documented: number of collected individuals, length, height, width, fresh weight, shell weight (as dry weight), tissue weight. The shell-somatic index – as a condition index for the mussels – is calculated using the above-mentioned parameters.

Eelpout or viviparous blenny (Zoarces viviparus) can be selected as a bioindicator for the marine environment. This is a very sedentary fish which spends its whole life in the same coastal area of the mud flats of the Wadden Zee. These fish can be caught in the gullies of the Wadden Zee mud flats using a "stake net", or at water level using a "dragnet". The biometric characterization is done on site and the following parameters are documented for each individual fish: fresh weight, length, sex, visible fish diseases, etc. The fish should be dissected if possible in a clean room (mobile laboratory); the liver and muscles are prepared and collected as target organs. Dissection is carried out using titanium blades or quartz glass knives to avoid contamination. Additional parameters (e.g. weight of liver and muscle) are documented during this preparational step. The target samples should be immediately frozen over liquid nitrogen on site.

Sea-bird eggs are often used as marine bio-indicators in monitoring programmes [18]. To collect large quantities of sea-bird eggs it is necessary to choose those birds that are sufficiently abundant so that the populations will not be disturbed or eliminated. For this reason *herring-gull eggs* (Larus argentatus) are often chosen for monitoring purposes. The herring-gull eggs are collected by ornithologists in bird sanctuaries, as depicted in Fig. 7.5 (areas for collecting birds/birds' eggs within the German ESB frame). Collecting is performed according to the standard operating procedures (SOP) of the German Environmental Specimen Bank [11] and the eggs are stored in a freezer until transport. The eggs are transported in transportable freezers at 4 °C. Further preparation is done according to corresponding guidelines (SOPs).

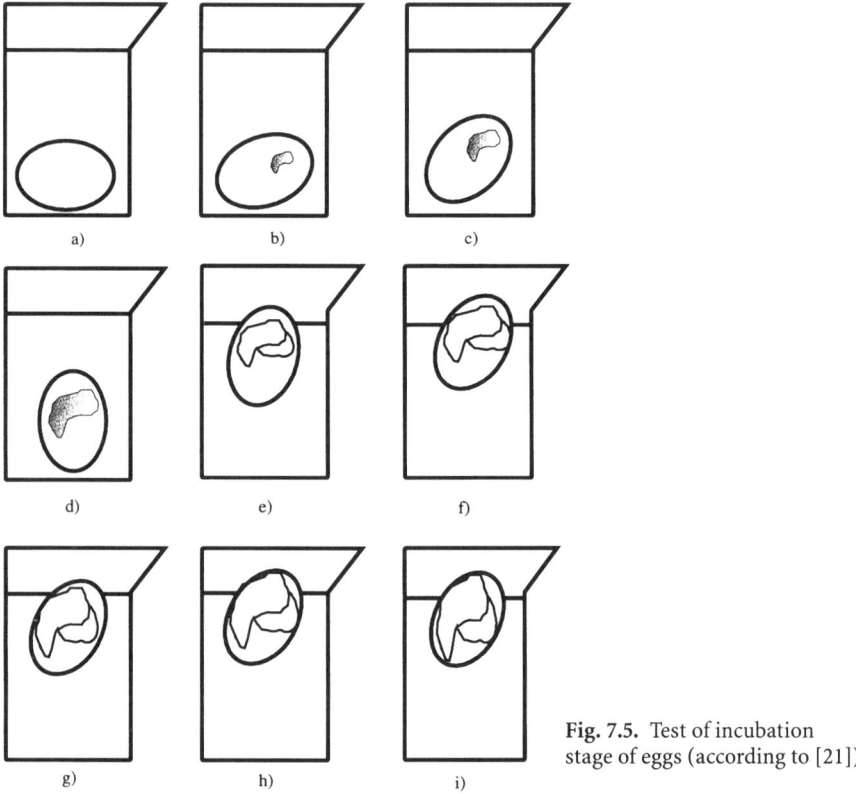

Fig. 7.5. Test of incubation stage of eggs (according to [21])

7.4
Sample Preparation

Monitoring programmes often use individual fish for their investigation aims. However, if the pollution of an area is being investigated, the natural variability of pollutants in the fish should be regarded and a homogenate of a suitable number of fish (number of individual fish has to be investigated by screening) should be prepared.

7.4.1
Preparation of Homogenized Sub-Samples

Directly after collection and biometric characterization of each fish, the material is pre-mixed and pre-crushed if necessary. Normally pre-crushing is done by pounding the material using a pestle made of titanium. During this step pre-mixing is performed simultaneously. The material from different containers is subsequently fed into a grinding device (e.g. a vibration mill, planetary ball mill, etc.) by a metering trough. For the purposes of the German ESB, a vibration mill has been specially constructed for cryogenic grinding [19]. The prototype of this vibration mill (CRYO-PALLA VMKT®) was made completely of stainless-steel. To avoid possible nickel and chromium contamination from the construction material, all parts of the grinding device coming into contact with the biological materials are made of titanium or PTFE. For the pleated bellows between the individual components (metering trough, grinding cylinder, outlet funnel) of the grinding device Teflon® PTFE is used.

Another grinding device for cryogenic grinding is a planetary ball mill with zirconium dioxide balls and mortars (Fritsch GmbH, FRG). Balls of different diameters and mortars with volumes of up to 500 ml are available. However, most of the volume is used by the grinding media (balls) so that only small portions of material can be ground.

The same is true of the grinding of lyophilized materials in agate mortars at room temperature (Fritsch GmbH, RETSCH GmbH, FRG); the volume of material that can be ground is limited by the volume of the balls and the volume of the mortars.

7.4.2
Cryogenic Grinding

Before starting the grinding process the whole grinding device (metering trough, mortars and balls, grinding cylinder and rods, etc.) is cooled with liquid nitrogen for several hours. If the temperature of the components is constant at less than $-190\,°C$, the grinding process is started without any further supply of liquid nitrogen. The deep-frozen sample material (also at $-190\,°C$) is filled into the grinding device and is ground to particle sizes of less than 200 µm in at least five discontinuous milling runs. The finely powdered sample material is collected in a container with a large volume, which is cooled from outside by liquid nitrogen.

The finely powdered material can then be bottled manually in scintillations glass vials using quartz-glass spoons. This type of glass vial, which is used for long-term storage in the environmental specimen bank, can store up to 10 g of homogenized material. The sub-samples are used for analytical characterization. Transportation to the analysts should be done in the gaseous phase over liquid nitrogen so that the cooling chain, starting at the collection site and ending at the analytical laboratory, is unbroken.

Complying with the cooling chain, it is possible to investigate even those substances that are not currently detected as potentially hazardous by the retrospective analysis of past samples.

7.4.2.1
Sediments

Up to now sediment samples (for the German ESB) have not been pretreated. The material is homogenized on site and sub-samples are directly frozen over liquid nitrogen. Until guidelines has been established for the collection of sediments, river sediment samples will be taken from surface sediments; marine sediment samples from mud-flats should be stored in the form of the total sediment fraction.

7.4.2.2
Bladderwrack

Bladder wrack (algae) should be collected at bimonthly intervals so that seasonal variations of contaminants can be investigated. The fronds of bladder wrack are deep frozen in the gaseous phase over liquid nitrogen directly after collection. The material is pre-crushed, and for seasonal variation studies of specific substances, small amounts are prepared for analytical characterization. The material collected from one year is pre-mixed and ground under cryogenic conditions as described above [9].

7.4.2.3
Common Mussel

Common mussels are collected bimonthly for the same reason as bladder wrack. The mussels are dissected under clean-room conditions at the frozen stage. The dissection of the soft-tissue parts is done under laminar flow (LF) boxes in an atmosphere of cold nitrogen gas. In this way only a few mussels at a time are taken from storage over liquid nitrogen and dissected; the other mussels remain in the gaseous phase over the liquid nitrogen until they have been pretreated. The soft tissue parts of the mussels are pre-crushed and pre-mixed using a titanium pestle. The shells of the mussels are discarded after some of the biometric parameters have been documented [20].

7.4.2.4
Eelpout

Fish such as eelpout, are dissected directly on site, preferably in a clean-room mobile laboratory. The fish are killed by a blow to the head or by cutting the neck using a pair of stainless-steel scissors. Each fish is individually identified. Biometric parameters (length, fresh weight, sex, etc.) are documented. Dissection is carried out under "clean-bench" conditions; the skin is removed, the muscle (filet) and the liver are collected as target organs for specific pollutants.

7.4.2.5
Herring-Gull (Egg)

The transportation of fresh herring-gull eggs is done in cooling boxes (electrical freezer or ice) at temperatures of about $+4\,°C$ to the laboratory (e.g. ESB facilities at Jülich). The incubation stage of each egg is documented. The eggs are placed in a glass of water according to Hayes' test [21]. The position of the egg in the water column shows the incubation stage (see Fig. 7.5). Only eggs according to Figs. 7.5 a–c are used for further pretreatment. The following parameters of each egg are documented: length, diameter, fresh weight. After opening both poles, the egg content (yolk and albumen) is blown out using inert gas (purified argon or nitrogen) into a container cooled with liquid nitrogen. The egg contents are pre-crushed and pre-mixed by a titanium pestle and ground to a fine powder under cryogenic conditions to avoid losses of volatile chemical compounds. After drying, the thickness of the shell of each egg is measured at several locations around the equator and the mean value is documented; the shell weight is also documented.

7.4.3
Bottling of Homogenized Standard Sub-Samples

For the German ESB, up to 250 sub-samples of homogenized material are stored for each collection site and each sampling period (the actual number depends more or less on the material available).

7.5
Conclusions

The increasing number of substances known to be potentially hazardous, toxic or environmentally relevant has led to an increasing demand for well-trained staff and methods with ultra-low detection limits. There is an additional lack of highly qualified laboratories, technicians, scientists, etc. in the field of collection, sample pretreatment and analytical characterization.

This continuously growing awareness of responsibility amongst the public and politicians and not least in industry has led to a multitude of measures, regulations, management controls, technical regulations, etc. (e. g. legislation on unleaded fuels, ban on DDT, ban on PCBs, drinking water regulations, etc.) A consequence of this expanding environmental policy is the continuously growing number of private and public laboratories. The equipment of these laboratories cannot be compared to those of 20 years ago. For this reason, a comparison of data from the past with data from sample characterizations today is not at present possible. That is one of the reasons for establishing the environmental specimen bank:

- because of their chemical stability, samples from the past, stored at cryogenic temperatures, can be used for comparison studies with today's samples [4, 5];
- ESB samples can be used as reference materials for quality assurance and quality control.

In 1991 alone, more than two million incorrect analytical results were produced, costing more than DM 10 million [22]. Results of analytical characterisation are often the basis for political decisions or are used by modern media for public information; therefore, their accuracy has to be verified by reference laboratories, which need exactly characterized samples.

After having performed every step – from sample collection, sample pretreatment, storage and analytical characterization – according to well-documented guidelines (standard operation procedures, SOPs) the evaluation and interpretation of all data is necessary. The evaluation and interpretation of the results of analytical characterization is the basis for:

- ascertaining the existing hazardous potential;
- ascertaining the scope of demand;

- recommendations and the establishment of measures with different priorities, e.g.
  - direct measures,
  - sanitation measures,
  - planning measures (e.g. restrictions on use).

However, other parameters also have to be taken into consideration:

- type and characteristics of the substance to be evaluated (e.g. stability, mobility, bioavailability, degradation and metabolism);
- type of binding of the substance to be evaluated.

Failure to consider specific parameters may lead to wrong interpretations. The evaluation and interpretation of analytical results is the most decisive aspect between analytical characterisation and establishing suitable measures and recommendations.

7.6 References

1. Lewis RA (1989) Forecasting, Assessment and the Nature of Real systems. What are the Limits? Environmental Monitoring and Assessment
2. GDCH/BUA (1987) Altstoffbeurteilung, p 32
3. SRU (Rat der Sachverständigen für Umweltfragen) (1987) Umweltgutachten 1987. Deutscher Bundestag, Drucksache 11/1569 and Kohlhammer, Stuttgart
4. BMFT (Bundesministerium für Forschung und Technologie) (1988) Umweltprobenbank – Bericht und Bewertung der Pilotphase. Springer, Berlin Heidelberg New York
5. Stoeppler M, Schladot JD, Dürbeck HW (1989) GIT Fachz Lab 10:1017–1020
6. Stoeppler M, Schladot JD, Dürbeck HW (1989) GIT Fachz Lab 11:1119–1124
7. Schladot JD, Stoeppler M, Kloster G, Schwuger MJ (1992) Analusis 20/3:M45–M49
8. FAO (Food and Agriculture Organization of the United Nations) (1976) Manual of Methods in Aquatic Environment Research, Part 2 – Guidelines for the Use of Biological Accumulators in Marine Pollution Monitoring. FAO Fisheries Technical Paper No. 150
9. Schladot JD, Backhaus FW (1988) Preparation of Sample Material for Environmental Specimen Banking Purposes – Milling and Homogenization at Cryogenic Temperatures. In: Wise SA, Zeisler R, Goldstein GM (eds) Progress in Environmental Specimen Banking. NBS Special Publication No. 740, 184–193
10. Schladot JD, Backhaus FW (1993) Probenahmerichtlinien für Blasentang (*Fucus vesiculosus*), Silbermöweneier (*Larus argentatus*) und entsprechende Richtlinien für die Aufarbeitung sämtlicher Umweltproben bei tiefen Temperaturen. Umweltbundesamt 1993 (in press)
11. Umweltbundesamt (ed) (1996) Standard Operating Procedures der Umweltprobenbank des Bundes. Erich Schmidt-Verlag (in press)
12. Helmers E: (cf. Chapter 4)
13. Mart L: personal communication
14. Ostapczuk P: personal communication
15. Kafadar K: personal communication
16. Backhaus F, Schladot JD (1993) Probenahmerichtlinie der Umweltprobenbank des Bundes für Miesmuscheln (*Mytilus edulis*). In: Konzeption der Umweltprobenbank – Fortschreibung und Ausbau, Bundesministerium für Umwelt, Naturschutz und Reaktorsicherheit. BMU Drucksache, Dec. 1993
17. Common Wadden Zee Secretary: personal communication
18. Becker P-H, Conrad B, Sperveslage H (1989) Chlororganische Verbindungen und Schwermetalle in weiblichen Silbermöwen (*Larus argentatus*) und ihren Eiern mit bekannter Legefolge. Die Vogelwarte 35:1–10

19. Schladot JD, Backhaus F, Reuter U (1985) Beiträge zur Umwelt Probenbank – I. Studie zur Probenhomogenisierung bei tiefen Temperaturen unter Berücksichtigung der für die Umweltprobenbank notwendigen Parameter. Jül-Spez-330, Forschungszentrum Jülich GmbH (KFA)
20. Schladot JD, Backhaus F (1992) The Common Mussel (*Mytilus edulis*) as Marine Bioindicator for the Environmental Specimen Bank of the Federal Republic of Germany. In: Roßbach M, Schladot JD, Ostapczuk P (eds) Specimen Banking. Springer, Berlin Heidelberg New York, p 75
21. Hays H, LeCroy M (1971) Wilson Bull 83:425–429
22. Tölg G (1993) personal communication

Chapter 8
Biological Samples

GERHARD WAGNER

8.1
Introduction: Ecological Basis – Information Content, Function and Indicative Value of Biological Specimens

An important task of environmental protection is to observe continuously the concentrations and effects of anthropogenically-inducted chemicals in the environment. The control of emissions by technical means and the measurement of deposition rates and concentrations of chemicals in different environmental media such as air, rainfall, soils, etc. are two important steps to fulfill this task. The necessary third step is the monitoring of concentrations of suspicious substances and their effects in biological objects (bioindicators), reaching from green plants (primary producers) through different steps of the nutrition pyramid up to humans. Results of this biological monitoring form a basis for the assessment and evaluation of possible environmental risks associated with chemicals. Biological samples taken from the environment should not only represent environmental conditions as accurately as possible, they should also possess probative value in the case of legal conflicts. However, for the sampling of biological specimens some specific scientific and technical requirements have to be considered; these will be discussed and explained in this chapter.

Biological samples from the environment are normally not taken as the actual targets of an investigation (as is the case e.g. for medical or forensic samples). These samples are mainly used and analysed as representatives of larger entities or similar or related environmental compartments. This requires the selection of standardised (bio)indicator systems, which react with known specificity and sensitivity to environmental chemicals and have the capability of spatial and/or temporal integration. Such indicator systems can be efficiently and reproducibly analysed and evaluated vicariously for the total entity of sensitive targets in the environment to be observed, targets which are often extremely variable with respect to space, time and physiology [see 1]. Bioindicator systems are preferably used as a supplement to technical-environmental monitoring systems in such cases, where potential integral effects of complex or unknown immission types have to be detected and quantified. Such effects may occur on different levels from specific organs of single organisms up to whole ecosystems. Bioindicators are also preferred in such cases, where they offer advantages due to their high sensitivity towards a broad spectrum of substances or because of their ability to accumulate a substance over an extended period of time or to integrate its influence in an area of known and relevant size. This is namely the case, if the sensitiveness of available analytical methods for dangerous substances is too low to find them in other environmental compartments such as air, water and soils. Bioindicators are also used as relatively cheap procedures for large-scale preliminary investigations to prepare and optimise the application of more specific and expensive methods such as analysis of air, deposition or soils in extended polluted areas [2].

In addition to the concentrations of toxic substances and their metabolites, biological specimens can also be analysed for essential components and a broad spectrum of possible biochemical, physiological, morphological and/or genetic effects. Organisms and biological communities normally do not react to single components or substances in their environment; rather they show the effects of the totality of all the acting substances and environmental factors. Decisive for the use of biological specimens is their ecotoxicological relevance, that means the relevance or indicative function that the effects observed have for other living organisms and communities including humans.

8.2
Quality Assurance in Biological Sampling

In biological sampling as in other fields the three major elements of quality assurance are adequate planning, documentation and control. For the planning phase a clear definition of the task, the area and the objectives of the investigation is needed. Results of the planning phase are: the exact description of the sampling area and any necessary subdivisions of it, the choice of adequate specimens, tools, materials, timing, statistical sampling designs, standardized written sampling methods (standard operating procedures, SOP), and agreement on a written specified sampling plan. Detailed, formalized and, as far as possible, on-line documentation of the whole process and the results is the basis for effective recognition and control of possible errors in the whole process. It is necessary to document completely and unambiguously: the timing and the results of the sampling; the existing weather conditions; the location, extent and essential properties of the sampling area; the composition and constitution of the samples; possible deviations from the SOP and the sampling plan and special events during the process; as well as all the steps of treatment, transportation and storage of the samples.

8.2.1
Representativeness of Biological Specimens and Samples

In general sampling means that a set of samples has to be obtained which is in the right form and quantity needed for analysis and which represents the total object to be analysed. Representative in this respect means that the sample has to show a maximum similarity to the total entity being studied. Therefore, the first requirement of quality assurance is to make sure that the samples really are what they stand for. The general requirements are the selection of suitable and representative sub-areas, specimen types, sampling methods and sound statistical designs. Risks of contamination or loss, natural variability, as well as the interfering influences of biotic and abiotic factors have to be taken into account. The following factors determine the representativeness and indicative value of the specimen type and the reproducibility of the sampling results [3, 4]:

- Exposition and accumulation behaviour of the selected specimen type;
- sampling period: time of the day and of the year, phenological stage of seasonal development and specific influences of weather conditions before and during sampling;
- inhomogeneity of the sampling site, especially with respect to edaphic and micro-climatological influences concerning immission and deposition of pollutants;
- the structure of the population to be sampled: abundance and composition with respect to age, sex or physiological condition, fluctuations or trends;
- selection of individuals or groups of the population with specific attributes as a selective effect of a sampling method;
- number of random samples; and last but not least;
- potentially disturbing biotic and abiotic influences.

Comprehensive knowledge of the influence exerted by the above factors is decisive for the final result. Such knowledge has to be gained as far as is necessary by preliminary examinations [5]. By laying down criteria for the selection, delimitation and standardization, a sufficient degree of reproducibility in the sampling can be achieved and analytical results that are quantitatively comparable can be expected.

8.2.2
Potential Errors in Biological Sampling

Contamination, loss, metabolism or any other alteration of the sample properties and composition are errors which have to be minimized; this implies a lot of general requirements. In the technical field, a lot of principles, rules and techniques have been worked out to optimize sampling quality; these are generally also important in environmental sampling. But there are some additional sampling problems to be solved when investigating specific sections of the environment, e.g. the contamination of a river and its fish population by xenobiotics [5–8]. The main problem for quality assurance in environmental sampling is how to guarantee the biological, ecological and geographical representativeness of the samples [4, 9–12]. The samples have to be chosen and taken so as to give relevant information about a specific area, period of time, environmental medium, species or population [13].

Errors in this case are all the possible deviations from the existing, but unknown true, concentration x of a substance in a given matrix of environmental origin and a specified location or area. This can mean (a) random errors of a few per mille or per cent (inaccuracy) of the result, (b) systematic errors due to contamination or loss of matrix components, (c) errors due to lack of representativeness of the sample, respectively (Fig. 8.1). True random errors can be recognized and partly eliminated by analysing a multitude of replicate samples. (Fig. 8.1, a and b). Unconsidered loss or increase of constituents of the matrix other than the substance being analysed (e.g. water or volatile metabolic products) can lead to systematic errors of + or − a few percent (Fig. 8.1 c). Typical systematic errors are also caused

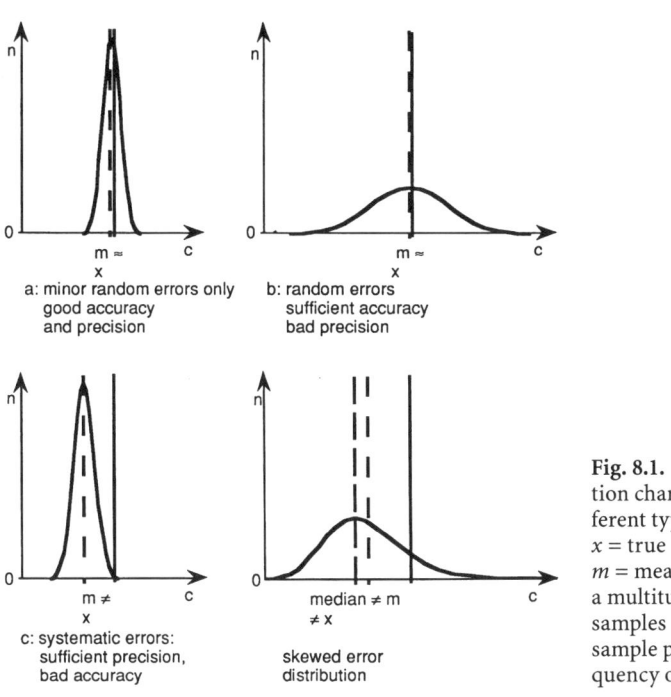

Fig. 8.1. Frequency distribution characteristics of different types of errors [14]. x = true concentration; m = mean concentration of a multitude of replicate samples or median of a sample population; n = frequency of occurrence

by contamination or loss of the sought-after substance during sample processing, storage or preparation (Fig. 8.1d).

Loss may lead to errors of up to -100% (e.g. total loss of volatile substances, absorption of lipophilic substances by plastic containers, adsorption or co-precipitation of ions, etc.), while gross systematic errors such as contamination can cause errors of more than $+100\%$, up to as much as several 100% or 1000% of the original concentration of the substance or even more. Similar errors occur if samples are not representative in a geographical, temporal, ecological or physiological sense. This can happen if, for example, a particularly highly contaminated spot of the area has been chosen for measurement or sampling, if the timing is bad or if an inappropriate indicator species has an exposure or accumulation behaviour that is different from that of the total quantity to be evaluated. Several of such systematic or random errors can overlay each other. This leds to skewed error distributions as shown in Fig. 8.1d. The effects of such gross errors cannot adequately be described by standard deviation; rather the logarithmic values will be normally distributet in such cases.

Table 1 demonstrates that random errors, which cause only more or less bad precision, are to be expected only during certain steps of sampling and sample preparation. Most of the other steps mainly involve risks of systematic and absolute (qualitative) errors, which detract from the trueness of the results and cannot be corrected or improved by increasing the number of random samples. Because of the different nature, behaviour, propagation, risks and consequences of these

Table 8.1. Characteristics, quality and risk of errors for the different steps of biological sampling

Phase Step	Main sources and characteristics of possible errors	Quality of errors	Risk of serious errors
Planning			
Definition and subdivision of the area	Spatial variability of ecological factors and contaminants, heterogeneity (hot spot or nugget effects)	Systematic + random	High
Selection of specimens	Ecological or physiological differences in exposure and accumulation behaviour	Systematic	Moderate
Stratification	Biological, physiological or spatial variability, inhomogeneous material	Systematic	Moderate
Sampling method	Statistical and physiological representativeness, skewed distribution, contamination or loss	Random and/or systematic	High, partly controllable
Number of samples	Too few replicates, pooling, statistical representativeness	Random	High
Sample mass	Statistical representativeness,	Random	Normally low
Timing	Temporal variability, phenology, weather conditions, phases in reproduction cycles	Random or systematic	High
Sampling			
Weather conditions	Unreproducible deposition, leaching or matrix effects	Systematic	Very high
Packaging	Contamination or extraction by tools and container material, volatilisation	Systematic	Moderate, controllable
Sample conservation during the sampling	Losses by metabolism, chemical ractions, volatilization or translocation (most serious for air filters, water and animal tissues)	Systematic or absolute	Moderate to very high
Transportation	Contamination or loss by volatilization, metabolism, separation (most serious for soil, sludge and water samples)	Systematic to absolute	High
Storage Short-term to long-term	Contamination or loss, metabolism, extraction, change of binding form or weight basis, speciation, solubility	Systematic to absolute	High to very high
Sample preparation			
Cleaning, washing	Contamination by the laboratory environment or loss by leaching	Systematic	High
Drying	Volatilization, oxidation, contamination	Systematic	Moderate
Homogenization	Contamination, disregard of skewed distribution	Systematic	High
Subsampling, aliquotation	Contamination, separation, heterogeneous particle and analyte distribution, statistical representativeness	Random	Moderate

various types of errors, they cannot compensate each other and may severely disturb any quantitative comparisons.

8.3
Development of Specified Sampling Plans

In order to obtain satisfactory, reproducible and quantitatively comparable results, which if required also have a probative value, the objectives, strategies and methodological details of an environmental analysis should be planned and layed down in a task-specific sampling plan, containing the following points (as a minimum):

- exact definition of the specimen type;
- description of the sampling or capturing method (reference to the guideline used and all the tools and containers needed;
- definition of the tempoal extent of the samples (sampling period);
- definition of the sampling area and if necessary, its subdivisions;
- definition of the type and density of the sampling grid and the sample-size necessary to achieve the required precision;
- definition of the date, period or rotation of the sampling (for biological specimens these dates should not only be defined by calendar dates but also by phenological data and atmospheric conditions);
- definition of a sampling design to make sure that the results are statistically sound.

These commitments have to be based on the results of preliminary studies and/or specific experience (see Sect. 8.3.2) and should form the basis of the working contract and quality control in the case of extended examinations. General instructions are given in the literature (e.g. [5]); more specific ones can also befound (e.g. [4, 7, 9, 10, 14, 16]. All the specifications have to be based on a clearly formulated description of the task.

8.3.1
Standardization of Sampling Procedures

Concentrations of major and trace elements and pollutants in biological matrices show a fairly high degree of biological, temporal and spatial variability on both small and large scales. This is true for both undisturbed ecosystems and those that have been anthropogenically influenced, managed or polluted. The concept of biological variability includes differences between species, between individuals of the same species, between different organs of the same individual (e.g. leaves, twigs, bark, wood, roots) as well as between analogous parts of an organism (e.g. leaves of different ages and growing positions in the crown of a tree [17, 18]. For comparative investigations this kind of variability has to be minimized; this is done by standardizing and stratifying the sampling so as to obtain samples which are comparable and reproducible over spatial and temporal distances. Standardization and stratification reduce the variety of the factors correlated with the sought-after parameters and their variability. However, although such samples do represent a well-defined part of the total entity or population rather precisely, they do not represent the whole of it. For examples see [7, 8, 16, 19–22].

In order to improve the comparability of environmental analysis by standardization of the sampling process, sampling guidelines are needed in the form of standard operating procedures (SOP). Such standardized procedures are also a prerequisite for establishing threshold values and other standards. If the results of an investigation are to be compared with such standards the respective SOP has to be strictly observed. The highest degree of standardization is to be found in the methods laid down by major organizations or as national or international standards or laws. In Germany, VDI guidelines (VDI = Verein Deutscher Ingenieure, Association of German Engineers) are available for the growing, exposing, sampling and analysis of standardized grass cultures [23] and lichens [24], and for the sampling of leaves and needles of trees in their natural habitat [25]. VDI guidelines for other plants and animals as bioindicators are in preparation. ISO-standards (ISO = International Standard Organization) for improving the sampling of soils are available [26].

Examples of standardized sampling guidelines and specified sampling plans have also been developed in the framework of the German environmental specimen banking programme; they have partly been printed as a guide for environmental-specimen banking [27] and will soon be published in English by the Research Centre in Juelich, Germany. Some particularly important aspects will be explained on the following pages.

8.3.2
Definition and Delimitation of the Sampling Area(s)

Biological specimens are always related to specific, relatively extended areas whereas technical measurements are normally related to points. However, environmental analysis never asks for attributes of a single spot but always for a given area. In order to make the necessary extrapolation from one or more points to the whole area in question, the problem of spatial representativeness always has to be solved. Therefore, the main prerequisite for efficient and reproducible environmental sampling is the selection and delimitation of representative and homogeneous sampling areas. For the sampling of biological specimens in the field, it is always necessary to examine first the ecological conditions with regard to soil, land use, depth of the ground water level, exposure, slope, location of (potential) sources of pollution, local climatic conditions, screening by vegetation or buildings, etc. If the area to be analysed is too heterogeneous, it has to be divided into sufficiently homogeneous parts. Inhomogeneous sampling areas or populations always increase the variation of the analytical results between different individual random samples and lead to skewed (often log-normal) distributions of the measurement results. However, for calculating the arithmetic mean (or for the pooling of samples) normally distributed values are always needed. Quantitative comparisons between the results obtained for different areas and/or sampling dates only make sense if, in addition to the (mean) values, the statistical variation is also known [5–7]. The homogeneity of the area and the population to be studied has to be tested by preliminary investigations (e.g.

Biological Samples

screenings). If this is not the case, the area has to be divided into homogeneous subareas; sometimes it may also be possible to reduce the area to be analysed down to the really contaminated core [2, 16]. A simple and often sufficient method of evaluating such preliminary results is to draw or plot a cross diagramm (see Fig. 8.2).

In the case of Fig. 8.2, the first two sampling points (0556 and 0557) clearly show high concentrations of the element barium. Without the need for mathematical models, the aggregation of data or the more difficult numeric procedures such as variance, standard error, etc., such a plot allows the level and distribution of concentrations as well as inhomogeneities or outlayers to be directly assessed. Such plots can also be constructed for more than one element [15]. Histograms may give more specific information about the form of distribution of the analytical results. The results for random samples taken from homogeneous sampling areas are always normally distributed (see. Fig. 8.3). This is a condition for the further use of parametric statistical methods. Inhomogeneous sampling areas often lead to skewed, often log-normal distributions which require non-parametric statistical methods.

8.3.3
Necessary Sample Size and Confidence Intervals

Confidence intervals support the objective evaluation of mean values of random tests by specifying a measure of uncertainty according to the span of the interval of true means related to a given level (a) of significance. That means that the true but unknown parameter with the probability (1-a) exists between the borders of the confidence interval. The width of the interval depends on the required level (a) of significance and the number of single random samples. For a normally distributed collection of random samples and a given level a of significance, the span of confidence as

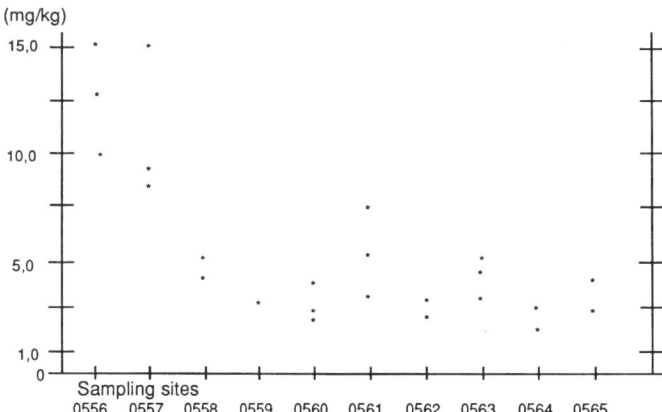

Fig. 8.2. Cross diagram of the concentrations of barium in spruce shoots from a sampling area in the Bavarian Forest National Park (from [15]). (Analysis: Research Centre Juelich, Institute for Applied Physical Chemistry)

well as the number of samples necessary to comply with a given width of the interval can be calculated for a mean m. The equation therefore is (according to [28]):

$$n = (4 * z^2 (a/2) * s_x^2)/KIB^2 \tag{8.1}$$

z = Z-value of the standardized normal distribution delimiting 1-a % of the area under the distribution curve
s_x = estimation of the standard deviation of the random samples
KIB = width of the confidence interval

Figures 8.2 and 8.3 and Table 8.2 show results of an investigation of one-year old spruce shoots taken from a sampling area of several square kilometres, sampled for the German environmental specimen banking programme in the Bavarian Forest National Park. From each of 10 randomly-selected sites, 3 spruce trees were sampled. The elemental analysis was made by the Institute of Applied Physical

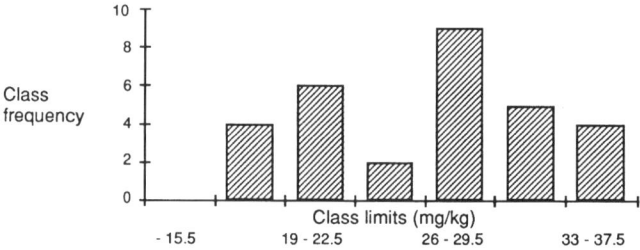

Fig. 8.3. Histogram of the concentrations of zinc in spruce shoots from a sampling area in the Bavarian Forest National Park (from [15]). (Analysis: Research Centre Juelich, Institute for Applied Physical Chemistry)

Table 8.2. Width of confidence intervals of selected substances (calculated from data of a screening of spruce shoots from the Bavarian Forest National Park, 1990 (from [15])

Substance	z-value	Mean (units/kg)	s	n	Half interval	
					In units/kg	In % of the mean
Copper	1.65	3.57 (mg)	0.68	30	0.204	5.6
Magnesium	1.65	809.53 (mg)	115.76	30	34.87	4.3
Barium[a]	1.65	5.83 (mg)	3.96	27	1.26	21.6
Barium[b]	1.65	4.03 (mg)	1.5	21	0.54	13.4
Cadmium	1.65	49.15 (mg)	23.4	30	7.05	14.3
Lead	1.65	2.0 (mg)	0.71	30	0.22	10.8
Benzo(a)pyrene	1.65	1.73 (mg)	0.65	30	0.2	11.4
Fluoranthene	1.65	10.18 (mg)	3.29	30	0.99	9.7
Pyrene	1.65	5.52 (mg)	2.01	30	0.61	11.0

[a] All samples, [b] without outlayers.

Table 8.3. Intervals and necessary sample size for 10% tolerance (from [15])

Substance	Half interval		n
	% deviation	In units/kg	
Copper	10	0.36 (mg)	10
Magnesium	10	80.95 (mg)	6
Barium	10	0.58 (mg)	126
Cadmium	10	4.92 (mg)	62
Lead	10	0.2 (mg)	34
Benzo(a)pyrene	10	0.17 (mg)	39
Fluoranthene	10	1.02 (mg)	29
Pyrene	10	0.55 (mg)	36

Chemistry, at the Research Centre in Juelich, the PAH were analysed by the Biochemical Institute for Environmental Carcinogens in Grosshansdorf and the statistical analysis was made by [15]. The data for barium are for demonstration only, because the distribution of barium concentrations is not normally distributed.

If the concentrations shall be allowed to deviate from the mean of the random test by maximum of 10%, the numbers of random samples given in Table 8.3 would be necessary.

The results demonstrate that only for some of the analysed substances the deviation limit can be reached with a reasonable effort. A compromise has therefore to be found, which means accepting higher deviations from the mean for highly dispersed substances such as cadmium and barium. If, for routine samplings, a maximum of 15 random samples per area is thought to be affordable, spans of confidence intervals shown in Table 8.4 are arrived at.

However, these exemplarily-determined numbers are specific for this case and may only provide a rough guide for investigations done in other regions or done using different methodological approaches, because in each case one has to deal with different patterns of dispersion and different influencing factors.

Table 8.4. Intervals and range of tolerance for $n = 15$ single samples (from [15])

Substance	n	Half interval	
		In units/kg	In % deviation
Copper	15	0.3 (mg)	8.1
Magnesium	15	66.34 (mg)	8.2
Barium[a]	15	0.69 (mg)	15.9
Cadmium	15	9.97 (mg)	20.3
Lead	15	0.3 (mg)	15.2
Benzo(a)pyrene	15	0.28 (mg)	16.0
Fluoranthene	15	1.4 (mg)	13.8
Pyrene	15	0.86 (mg)	15.5

[a] Without outlayers.

8.3.4
Sampling Time and Age of the Material

The time of day is of significance for the reproducibility of plant-materials sampling, because the changing activity of photosythesis during the day also causes changes in the biomass, water content and composition of the material to be sampled. These changes affect the fresh weight and the dry weight of leaves and may thus alter the results of elemental analyses in the order of a few percent of the mean value.

Much greater effects are caused by seasonal differences during the year. A lot of publications on the seasonal changes of nutritional and trace elements in plants are available. Wyttenbach and Tobler [18] analysed 20 elements in spruce needles and found three different groups with characteristic seasonal concentration values. Ahrens [29] analysed leaves and needles of different forest trees and found the highest concentrations during the first phase of leaf development for the essential trace elements copper, zinc and molybdenum. In the course of further leaf de-

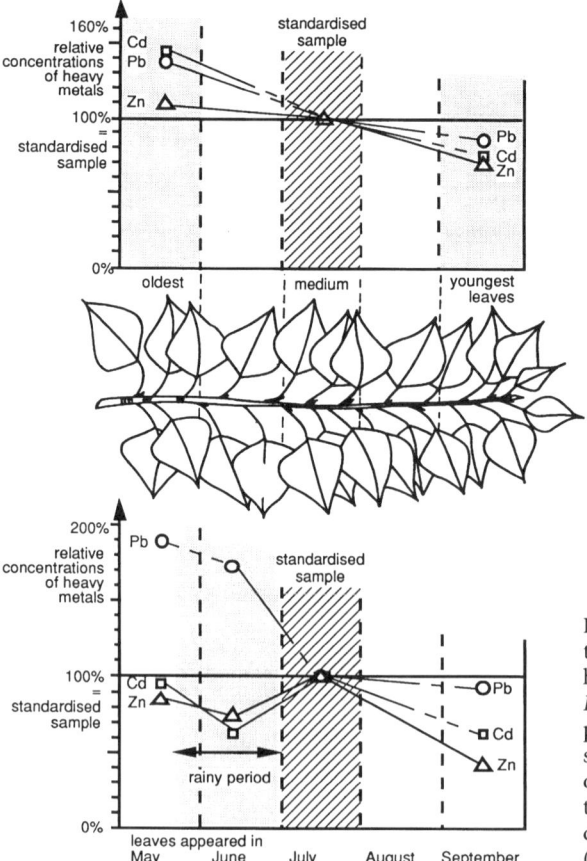

Fig. 8.4. *Above:* Influence of the age of poplar leaves on heavy-metal concentrations. *Below:* Influence of a rainy period in connection with strong infestation with aphids on the heavy-metal concentrations in leaves which had developed during this period (from [20])

velopment, the concentrations of these elements decline rapidly (correlated with the development of biomass) and then remain rather constant during the vegetation period. The total contents per leaf, however, increase continuously with increasing biomass. The concentrations of elements which are available in excess from the soil or from atmospheric deposition (e.g. lead) increase during the whole vegetation period. The trend may be interrupted by rainy periods by washing off and/or leaching (see Fig. 8.4). When the leaves begin to senesce in autumn, the more mobile nutrients such as nitrogen and potassium, often also trace elements such as zinc, iron and copper are resorbed by the plant and their concentration in the leaves decreases; the concentration of the less mobile elements such as calcium and magnesium, together with the surplus and the potentially toxic elements, may increase rapidly by passive accumulation because biomass is metabolically reduced (see Fig. 8.5 and [17]). The concentrations of such elements also tend to increase from year to year in long-living plant organs, e.g. in spruce needles [30–32].

For the sampling of biological specimens it is principally important to find and define periods characterized by a minimum of internal or external changes, when the composition of the material to be sampled is as constant as possible. The sampling of aquatic organisms, for example, should not be done during the spawning period, because the rapid changes in the biomass of the organisms and their composition would lead to results that can hardly be reproduced or interpreted [e.g. 7, 33]. Weather conditions before and during the sampling are also factors which may have a strong influence on the sample composition, particularly for terrestrial plants. For temporal trend analysis in particular, it is important that the sampling dates or periods are not only fixed according to the calendar but also with an eye to atmospheric conditions and the phenological stage of vegetation development.

The variation of seasonal atmospheric conditions from year to year of course also influences the concentration in the ambient air, the processes of dry and wet

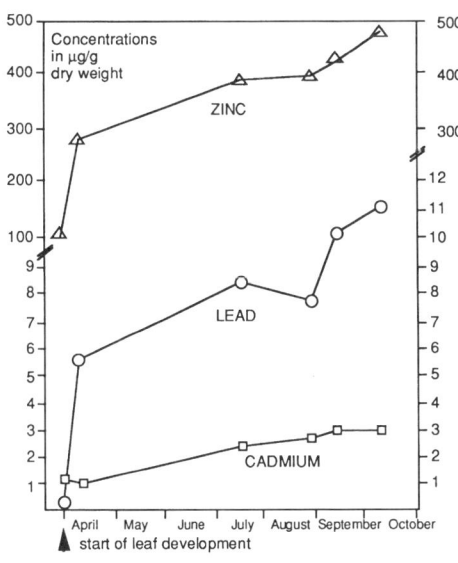

Fig. 8.5. Trends of heavy-metal concentrations in poplar leaves during the vegetation period (from [20])

deposition, the biological availability to organisms of pollutants and the physiological susceptibility of organisms. This is the main reason why temporal trends established by using bioindicators can give a more realistic picture of the effective pollutant load of an area than physical measurements of atmospheric concentrations and/or deposition rates are able to.

8.4
Specific Problems of Sampling and Preparation of Biological Environmental Specimens

8.4.1
Difficulties of Sampling Caused by Abiotic and Biotic Factors

It is not only the abundance or availability of the target organisms but also their susceptibility and accumulation behaviour in the presence of environmental chemicals that are influenced by changing abiotic factors such as atmospheric conditions, water and nutrient supply, singular events (e.g. storms) and attack by diseases, parasites or predators. The possibility of sampling a species at a certain place and time and the reproducibility of the sampling may thus be adversely affected. Many animal species respond to periods of optimal conditions with an exponential growth of their population. Because they are conspicuous and their available biomass is sometimes very high, such species are often considered to be particularly suitable as bioindicators, e.g. many small mammalia, insects and slugs. But particularly these species are characterized by the fact that during unfavourable periods the biomass and/or the number of individuals necessary for representative sampling is hardly obtainable; moreover changing diet during such periods gives rise to samples that cannot be well compared [12, 30]. The possible effects of these disturbing natural factors have to be considered when selecting specimens and drawing up a sampling plan. Of course the regeneration capacity of a population also has to be taken into consideration so as to avoid damage to their stock by sampling.

8.4.2
Semi-Experimental Solutions of Sampling Problems

An especially elegant solution to such problems is shown by the example of the zebra mussel (*Dreissena polymorpha*). This species has been chosen for environmental specimen banking as an ideal bioindicator for many fresh water ecosystems. However, regular sampling of the normally-abundant wild populations often failed because of disturbance caused by different biotic and abiotic factors: for example when there is high water, when the mussels became covered with algae or mud after storms, or when there is a high fluctuation in population density as a result of diseases or an attack by predators such as flocks of mussel-eating birds or fish. A method had to be found to improve the comparability of regularly-repeated sampling and to make sure that a sufficient amount of sample material is always available and easy accessible. For this purpose, plates of additive-free polyethene material have been built up into racks and placed in the water in order to be

Fig. 8.6. Racks made of non-contaminating materials ready for exposure in Lake Constance and colonization by zebra mussels (*Dreissena polymorpha*)

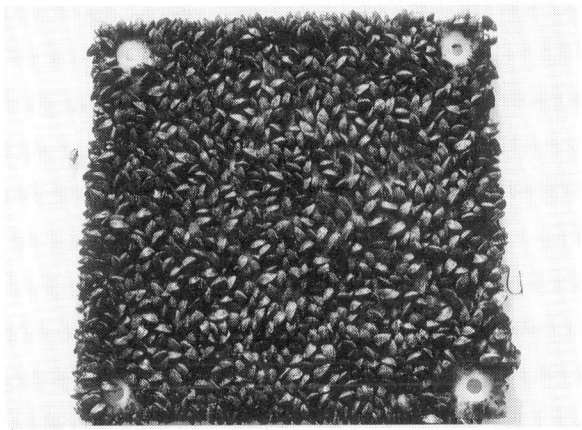

Fig. 8.7. Polyethene plate colonized by zebra mussels (*Dreissena polymorpha*) after one year

colonized by the zebra mussels (see Fig. 8.6). After colonization (Fig. 8.7), the racks are easily accessible for reproducibly sampling mussels of known age without contamination [3, 12]. This method eliminates most of the above-mentioned problems of disturbance.

Similar possibilities for improving and standardizing sampling procedures are also available for other animals [3, 16, 34]. In particular, honey-bees and their products such as honey, beeswax and propolis can easily be used as standardized biological indicators by the carefully-directed exposure of suitable beehives in the area(s) to be monitored [35, 36]. Another example is the standardized sampling of eggs or feathers from feral pigeons (and possibly some other bird species) by offering suitable carefully-directed breeding-places. Traditionally-used or carefully-directed breeding houses for feral pigeons for example can easily be used for the reproducible sampling of pigeon eggs for long-term trend monitoring (see Fig. 8.8 and [37]).

Fig. 8.8. Breeding house for feral pigeons in Bayreuth, Germany

8.4.3
Contamination or Deterioration of Plant Samples: the Problem of Washing

There are general problems of possible sample contamination (or loss or alteration of important constituents) during the processes of sampling, sample preparation, transportation and storage through contact with tools, containers and the environment of the working place. These problems are common to all environmental samples and will be discussed in more detail in other chapters of this book (especially 1 – 3, 6 and 9). However, some of these problems, occurring specifically within biological sampling, will be explained in the following by a few examples.

The accumulation and effects of pollutants in plants represent the effective input of those substances during the whole period of exposure. The degree of accumulation of a substance in a plant depends upon its bioavailability under the existing environmental conditions, its persistence, its access to the plant via air, soil or by precipitation and on the accumulation capacity of the plant itself [see 38 – 40]. The question as to whether the parts of the pollutant adhering to the plant's surface should be included in the analysis or not has to be answered according to the specific aims of the investigation.

The washing of plant samples may be necessary for the following reasons:
- to eliminate weather-dependent differences in surface contamination;
- to simulate the cleaning effects that usually occur during the preparation of food plants in the kitchen;

- to differentiate between differently-bound or located compartments of the total content of a contaminant in the sample material.

Specific problems are caused by the fact that particles on a plant surface can be bound very strongly by wrinkled or hairy structures or by becoming embedded in the waxy layers of a cuticle. Washing with water will only remove a relatively small proportion of these particles. More drastic washing procedures using aqueous solutions that contain surfactants, acids or complexing agents, or procedures aided by ultrasonic power, may be able to remove most or nearly all of these particles; on the other hand, biological membranes or cell walls may be damaged, thus causing relatively extensive leaching of substances from the inner parts of the plant. Removal of all the particles which are bound at the plant surface by epicuticular waxes can be achieved by using fat-dissolving agents such as chloroform [39], butyl alcohol or a mixture of toluene and tetrahydrofuran [31, 38, 41]. Such procedures, however, lead to a fragmentation of the matrix to be analysed, and this has to be considered in the calculation and evaluation of the results. New contamination of the sample is possible from the washing solution itself or from exposing the clean, wet and absorptive material to the laboratory atmosphere or potentially contaminating surfaces, especially during the subsequent drying step. When the drying is done in an oven with circulating air, it is often advisable not to dry the samples in open vessels or crucibles but rather inside closed parchment bags, which protect the samples from contamination by particles while at the same time being permeable to water vapour.

The question of selecting a suitable washing procedure, should it be really necessary at all, is rather critical and can only be answered according to the aims of the whole investigation. If, for example, the aim of the investigation is to evaluate the physiologically active concentrations in plants or the contamination of food plants or vegetables by environmental chemicals, washing of the material is absolutely necessary. The washing procedure chosen has to achieve the desired effect but should be as gentle as possible so as to avoid additional losses. Recommendations for washing procedures can be found in [38–43]. It is generally essential to test the effectiveness and reproducibility to the procedure used, because washing procedures always carry the risks of irreproducible effects, contamination and losses and they often spoil the samples rather than helping towards results (see Fig. 8.9).

If plant material is analysed as a representative of fodder or used as a passive sample of airborne dust, washing of the samples is inappropriate. In such cases mechanical abrasion during the sampling procedure should also be minimized and the weather conditions before and during the sampling need to be considered and recorded. During the sampling, every unnecessary contact with the material to be sampled and processed should be avoided. Tree leaves for example should generally be cut off (in most cases only the blades without stalks) while still in the field using clean scissors and in such a way that the leaves fall directly into the prepared sample container without touching anything else. This work should not be done in the laboratory in order to avoid specific laboratory contamination [see 20, 22, 43, 44].

A weather-dependent source of disturbance is also the contamination by particles of dust and soil which are whirled up by wind and strong rainfall and adhere

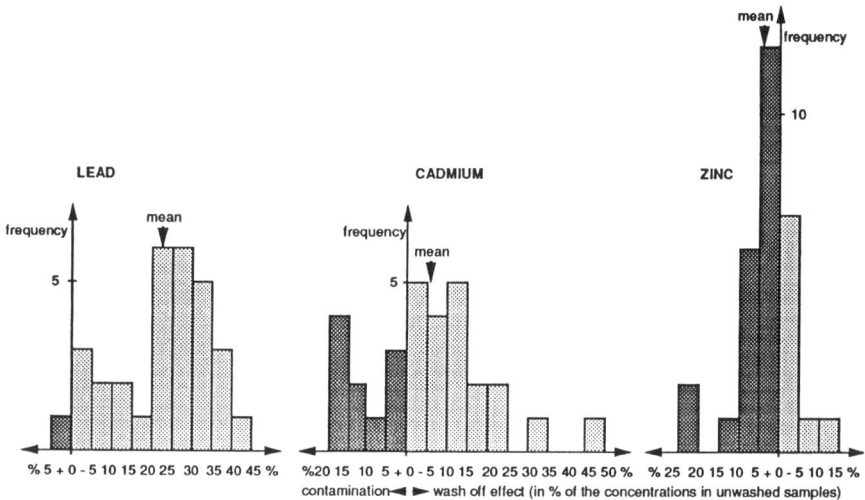

Fig. 8.9. Effects of a simple washing procedure (2× tap water + 1× de-ionized water, according to [23]. Removable parts of heavy-metal concentrations in poplar leaves (dark areas = contamination, from [20])

to the surfaces of plants growing near the ground. This problem can only be avoided by using higher-growing plants and by sampling their leaves at a sufficient, definite height above the ground. This has two advantages. The first is to avoid the extremely variable influences of the air layer adjacent to the soil. The second is to use the distributive movements of the air above the ground so that one sample is able to represent a fairly large area [20].

8.4.4
Principles Governing the Sampling and Preparation of Animal Tissues for Environmental Analysis

A precondition for using animals as bioindicators is the availability of a practicable sampling or capturing method which does not contaminate or alter the sample in any way [3]. Capturing methods using bites, poisons, preservatives, etc. are not eligible in most cases because of contamination and some other problems. Physical methods used to capture or kill animals for environmental analysis have also to be tested for contamination risks, e.g. by small fragments of bullets. Conformity with animal protection laws is also in important criterion. If animals are killed in traps and remain there too long, metabolic or decomposing processes can dramatically change the concentrations and/or distribution of chemicals in their bodies as well as the total biomass. Easily metabolized organic chemicals may be excreted, lost either by the enzymatic decomposition that occurs as a result of stress before death or by endolytic reactions after death [see 4]. The redistribution and excretion of contaminants could be demonstrated during the removal of the gut content from earthworms [45]. On the other hand it is necessary in many cases to distinguish between substances which are incorporated in body tissues (physiologically active) from those simply pass through the intestines of the organism (inside the body but physiological inactive). Therefore, the gut content or the entire gastrointestinal tract has to be separated.

To obtain satisfactory analytical results for larger animals, dissection is normally unavoidable. In vertebrates, for example the liver and kidneys are often considered to be organs where most of the burden of unwanted chemicals has accumulated. For specific elements or substances blood, fatty tissues or other organs may also be of special interest. In such cases the analysis of small amounts or concentrations of physiologically-active substances in these organs is the leading objective. However, these substances may occur in much higher concentrations in particles of the laboratory atmosphere; therefore contamination by dust, aerosols, vapours, tools, containers or chemicals in the laboratory is a severe problem since the samples to be dissected, prepared and analysed are small, supple, moist and often colder than the surrounding. To avoid airborne laboratory contamination, a clean bench is necessary for such work. If semivolatile organic compounds are to be analysed, the clean bench should additionally be equipped with an absorbing filter of charcoal.

To summarize, it is generally recommended that the manipulation of the sampled material be reduced to the absolute minimum that is necessary to reach the goal. In order to reduce laboratory contamination, preparatory steps should be done directly in the field as far as appropriate or under clean-air conditions in the laboratory.

8.5
References

1. Wagner G (1993) Einsatzstrategien und Meßnetze für die Bioindikation im Umweltmonitoring. In: Ries L, Wagner G, Fiedler H, Hutzinger O (eds) ECOINFORMA '92 Bd 4: Biomonitoring und Umweltprobenbanken, Umweltdatenbanken und Informationssysteme. Ecoinforma-Press, Bayreuth, p 23
2. Wagner G (1993) Large-Scale Screening of Heavy Metal Burdens in Higher Plants. In: Markert B (ed) Plants as Biomonitors for Heavy Metal pollution of Terrestrial Environment. Verlag Chemie, Weinheim New York, p 425
3. Müller P, Wagner G (1985) Untersuchung von Probenarten und Entwicklung von Probenahmerichtlinien für Biomonitoring im Rahmen der Umweltprobenbank. Umweltforschungsplan des Bundesministerium des Innern, Forschungsbericht i. A. des Umweltbundesamtes, Saarbrücken
4. Lewis RA (1987) Guidelines for Environmental Specimen Banking with Special Reference to the Federal Republic of Germany: Ecological and Managerial Aspects. U.S. MAB Report No 12
5. Green RH (1979) Sampling Design and Statistical Methods for Environmental Biologists. Wiley, New York, p 257
6. Keith LH (ed) (1987) Principles for Environmental Sampling. ACS Professional Reference Book, American Chemical Society, p 45
7. Bingert A, Göthberg A, Jensen S, Litzen K, Odsjö T, Olsson M, Reutergardh L (1993) The Science of the Total Environment 128:121–139
8. Ernst WHO (1994) Sampling of Plants for Environmental Trace Analysis in Terrestrial, Semiterrestrial and Aquatic Environments. In: Markert B (ed) Environmental Sampling for Trace Analysis. VCH Weinheim, pp 381–394
9. Paulus M, Klein R, Wagner G, Müller P (1994) Biomonitoring und Umweltprobenbank I: Das ökologische Rahmenkonzept zur Qualitätssicherung in der Umweltprobenbank des Bundes. -UWSF-Z Umweltchem Ökotox 6(4):223–231

10. Paulus M, Altmeyer M, Klein R, Hildebrandt A, Ostapczuk P, Oxynos K (1994) Biomonitoring und Umweltprobenbank II: Aufbau flächenrepräsentativer Probenahmen von Umweltproben zur Schadstoffanalytik am Beispiel der Regenwürmer in landwirtschaftlich genutzten Räumen. -UWSF-Z Umweltchem Ökotox 6(6):375–383
11. Paulus M, Klein R, Wagner G (1996) Biomonitoring und Umweltprobenbank VI: Anwendung von UPB-Strategien zur Qualitätssicherung in der Umweltüberwachung -UWSF-Z Umweltchem Ökotox 7
12. Wagner G, Klein R (1995) Sampling strategy in environmental monitoring of biological specimens. In: Quevauviller Ph (ed) Quality Assurance in Environmental Monitoring – Sampling and Sample Pretreatment, VCH Weinheim, pp 25–50
13. Koch R (1992) Environmental Analysis: How can results be interpreted? Fresenius Environ Bull 1992(1):682–687
14. Wagner G (1995) Basic approaches and methods for quality assurance and quality control in sample collection and storage for environmental monitoring. The Science of the total Environment 176:63–71
15. Fischer P (1991) Statistische Überprüfung und Optimierung von Beprobungsplänen für die Umweltprobenbank am Beispiel von Fichte (*Picea abies* L.) und Buche (*Fagus sylvatica* L.). Thesis, Biogeography, University Saarbrücken
16. Klein R, Paulus M (eds) (1995) Umweltproben für die Schadstoffanalytik im Biomonitoring – Standards zur Qualitätssicherung bis zum Laboreingang. – Gustav Fischer Verlag, Jena
17. Wagner G (1990) Variability of element concentrations in tree leaves on sampling parameters. In: Lieth H, Markert B (eds) Element Concentration Cadasters in Ecosystems (ECCE):41–54. VCH Weinheim
18. Wyttenbach A, Tobler L, Bajo S (1988) Trees 2:52–64, 204–212
19. Knabe W (1983) Immissionsökologische Waldzustandserfassung in Nordrhein-Westfalen (IWE 1979). Forschung und Beratung C 37, Reihe C, Landwirtschaftsverlag Münster
20. Wagner G (1987) Entwicklung einer Methode zur großräumigen Überwachung der Umweltkontamination mitels standardisierter Pappelblattproben von Pyramidenpappeln (*Populus nigra* "Italica") am Beispiel von Blei, Cadmium und Zink. In: Stoeppler M, Dürbeck HW (Hrsg) Beiträge zur Umweltprobenbank 5, KFA Jülich
21. Rahlenbeck SI (1991) Use of Plants in Monitoring Air Quality. WHO Air Hygiene Report 5, Federal Health Office/Bundesgesundheitsamt/WaBoLu, Berlin
22. Wagner G (1993) Plants and soils as specimen types from terrestrial ecosystems in the environmental specimen banking program of the Federal Republic of Germany. In: Stoeppler M, Zeisler R (eds) Biological Environmental Specimen Banking. The Science of the Total Environment 139/140:213–224
23. VDI (1978, 1980 and 1986) Response Dose Determination; Standardized Grass-exposure Method. VDI 3792 Parts 1–3 – Beuth Verlag, Berlin
24. VDI (1991) Measurement of Immission Effects – Measurement and Evaluation of Phytotoxic Effects of Ambient Air Polutants (Immissions) with Lichens, Method of Standardized Lichen Exposure. – VDI 3799 Part 2, – Beuth Verlag, Berlin
25. VDI (1991) Response Dose Determination, Standardization of Sampling of Leaves and Needles from Trees at their Natural Site. VDI 3792 Part 5, – Beuth Verlag, Berlin
26. ISO International Standard Organization (1994) ISO/CD 10381, parts 1–5: Soil Quality – Sampling
27. UBA (Umweltbundesamt/Federal Environmental Agency) 1989: Guide for Environmental Specimen Banking of the Federal Republic of Germany. Working Document compiled for the US/German Seminar on "Environmental Specimen Banking" in Ottawa, 1989
28. Bortz J (1985) Lehrbuch der Statistik für Sozialwissenschaftler. Springer Verlag, Berlin Heidelberg New York Tokyo, 2. Auf.
29. Ahrens E (1964) Allg Forst- u. Jagdzeitg 135:8–16
30. Müller P, Wagner G (1986) Probenahme und genetische Vergleichbarkeit (Probendefinition) von repräsentativen Umweltproben im Rahmen des Umweltprobenbank-Pilotprojekts. BMFT-FB-T 86-040

31. Wyttenbach A, Schleppi P, Bucher J, Furrer V, Tobler L (1994) The Accumulation of the Rare Earth Elements and of Scandium in Successive Needle Age Classes of Norway Spruce. Biological Trace Element Research 41:13-29
32. Wyttenbach A, Schleppi P, Tobler L, Bajo J, Bucher J (1995) Concentrations of nutritional and trace elements in needles of Norway spruce (*Picea abies* [L] Karst) as functions of the needle age class. Plant and Soil 168-169:305-312
33. Stöcker G (1993) Analytical and biological data of banked samples - a requirement to interpret accumulation and leaching of pollutants in biological specimens. In: Stoeppler M, Zeissler R (eds) Biological Environmental Specimen Banking. Sci Total Environ 139/140:491-505
34. Klein R (1993) The animal specimens of terrestrial and limnetic ecosystems in the Environmental Specimen Banking Programme of Germany. The Science of the Total Environment 139/140:203-212
35. Bromenshenk JJ, Carlson SR, Simpson JC, Thomas JM (1985) Pollution monitoring of Puget Sound with Honey Bees. Science 227:632-634
36. Höffel I, Müller P (1983) Schwermetallrückstände in Honigbienen (*Apis mellifica* L) in einem Ökosystem (Saarbrücken). Forum Städte-Hygiene 34:191-193
37. Altmeyer M (1993) Biomonitoring mit Stadttaubeneiern zur Erfassung von Chemikalien und deren Wirkungen in Verdichtungsräumen. Diss Universität des Saarlandes, Saarbrücken
38. Wyttenbach A, Tobler L, Bajo S (1989) Toxicological and Environmental Chemistry 19:25-33
39. Wyttenbach A, Bajo S, Tobler L, Keller T (1985) Major and trace element concentrations in needles of Picea abies: Levels, distribution functions, correlations and environmental influences. Plant and Soil 85:313-325
40. Ernst WHO (1990) Element (re)translocation in plants and its impact on representative sampling. In: Lieth H, Markert B (eds) (1990) Element Concentration Cadasters in Ecosystems (ECCE):17-40. VCH Weinheim
41. Schwedt G, Jahns G (1987) Fresenius Z Anal Chem 328:85
42. Bundesgesundheitsamt (1979) Bekanntmachungen des BGA: Probenvorbereitungsverfahren für die Bestimmung von Schwermetallen in und auf Lebensmitteln. Bundesgesundheitsblatt 22/15:277-279
43. Markert B (1995) Quality Assurance of Plant Sampling and Storage. In: Quevauviller P (ed) Quality Assurance in Environmental Monitoring for Trace Element Determination. Weinheim
44. Wagner G (1993) Umweltprobenbanken - neue Instrumente für Umweltforschung, -analytik und -planung. In: Ries L, Wagner G, Fiedler H, Hutzinger O (Hrsg) ECOINFORMA '92 Bd 4: Biomonitoring und Umweltprobenbanken, Umweltdatenbanken & Informationssysteme, -Ecoinforma-Press, Bayreuth:71-80
45. Riss B, Müller P (1989) Ökologische und rückstandsanalytische Untersuchungen zur Eignungsprüfung von Regenwurmarten als Indikatororganismen für die Umweltprobenbank. UFOPlan des BMU Umweltplanung/Ökologie i. A. des UBA, Saarbrücken

Chapter 9
Sampling of Industrial Material (Sampling for the Balancing of Elements in the Cement Industry)

WOLFRAM RECHENBERG and GEORG BACHMANN

9.1
Introduction

The behaviour of various elements of environmental interest in an industrial process is normally investigated by balancing the input and output of each element. In this way the behavior of several minor constituents and trace elements in the cement clinker burning process [1–8] and in power generating plants have already been investigated [9–16]. The results of investigations on cement kilns have led to a remarkable reduction of the emission of some of these elements. This chapter reports on the experience gained with cement kilns; a great deal of this may be transferred to other industrial processes.

9.2
The Cement Clinker Burning Process

9.2.1
Generalities

The cement manufacturing process can be divided into 3 parts, which are shown in Fig. 9.1 [17]. The upper part shows how the blasted raw material coming from the quarry is crushed and homogenized in blending beds, which will normally hold approximately 100 000 tons. They are built up from above and from the side of the pyramid that forms the material for the raw mill is taken. The milling process produces a dust, the raw meal, which is quantitatively driven by a waste-gas stream to an electrostatic precipitator. The precipitated solids are carried over to clinker silos that comprise the first part of the clinker burning section, as shown in the middle part of the figure. The kiln feed is again homogenized and transferred to the preheater between the first and second cyclone. The dust is suspended in the hot gas, precipitated in the first cyclone and again suspended in the hot gas between the second and third cyclone. The precipitations and suspensions are repeated until the partly-precalcined material enters the kiln. The gas in the kiln inlet has a temperature of about 1000 °C, the solid material of about 800 °C. The material further approaches the flame where it gains a temperature of about 1500 °C while sintering to clinker. The clinker is cooled and stored in the clinker silo.

The third part of the figure, below, shows the cement mill, in which the clinker is milled together with gypsum to form Portland cement. Sometimes blast furnace slag, pozzolana or fly ash are added to the mill to obtain slag cement, pozzolana

Sampling of Industrial Material

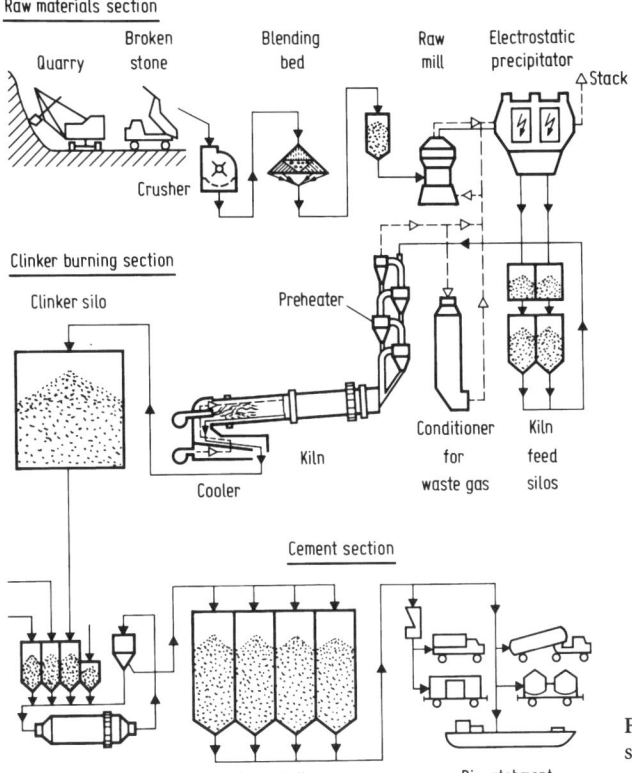

Fig. 9.1. Schematic presentation of the cement making process [17]

cement or fly ash cement. Finally the cements are stored in silos and later shipped on request.

With regard to environmental protection only the clinker burning process in the middle part and the raw mill and the electrostatic precipitator in the upper part are of interest. The boundaries of an external balance are therefore the inputs of raw material stream entering the raw mill and the fuel entering the kiln. The outputs are the waste gas from the electrostatic precipitator and the clinker [18].

9.2.2
Balancing

If the emissions from an industrial process are to be minimized, the inputs and outputs of the element in question have to be balanced. The flow of non-volatile substances, such as calcium oxide and alumina, should be balanced without any deficits. In practice, however, there are always balance deficits, which result from inevitable errors in sampling, sample preparation and analysis. The balance deficit is a measure of how good the balance is (Chap. 9.5).

The exhaust gases from the kiln pass through the preheater from below in counterflow to the kiln feed. Besides gaseous components from the combustion and dissociation of the feed material, the hot gas contains feed-material dust as well as a series of other gaseous and vaporous compounds. These include alkalis, chlorides and sulphur-bearing compounds, formed from the evaporated and dissociated components of the feed material and the fuel as a result of reactions in the gas phase. In addition the gas contains dust. The evaporated compounds condense on the dust while cooling inside the preheater. Subsequently, the gas-dust mixture is passed back into the hot kiln, thereby creating internal circulation inside the preheater. The dust in the crude gas leaving the preheater is precipitated in the electrostatic precipitator and then combined with the incoming raw materials to give the kiln feed, thus creating external circulation. Through the influence of internal and external circulation, the deficits of the external balance may increase. For a complete description of the behaviour of any particular element, an internal balance is required. A comparison of the inputs and outputs of a preheater is called an internal balance. In Table 1 the materials for both balances are listed [12, 14, 18].

The flowing materials listed in Table 1 have to be measured over a sufficiently long period of time, sampled and analysed [12, 14, 18].

9.2.3
Preheater Systems

The two most commonly used preheater systems are completely different: the cyclone and the grate preheater [19]. Only the cyclone preheater kiln will be referred to in this chapter; it is shown schematically in Fig. 9.2.

The exhaust gases from the kiln pass through the cyclones in counterflow to the kiln feed. The kiln feed enters the cyclone preheater with a temperature of approximately 60 °C; it is suspended, then precipitated, and again suspended until it reaches the kiln in a partly calcinated condition. In contrast the hot gas from the kiln has a temperature of approx. 1000 °C, but when it leaves the cyclone preheater it has cooled down to about 400 °C. After that there are two possibilities. With the evaporation cooler in an on-mode, the crude gas is passed

Table 9.1. Flowing masses of external and internal balances [6, 8, 18]

Type of balance	External balance	Internal balance
Inputs	Raw material, untreated	Kiln feed
	Fuels, untreated	Fuels, ready for use
Outputs	Clinker	Clinker
	Clinker dust	Clinker dust
		Crude gas (dust & gas)
	Removed dust	Removed dust
	Emitted dust	Emitted dust
	Gaseous emissions	Gaseous emissions

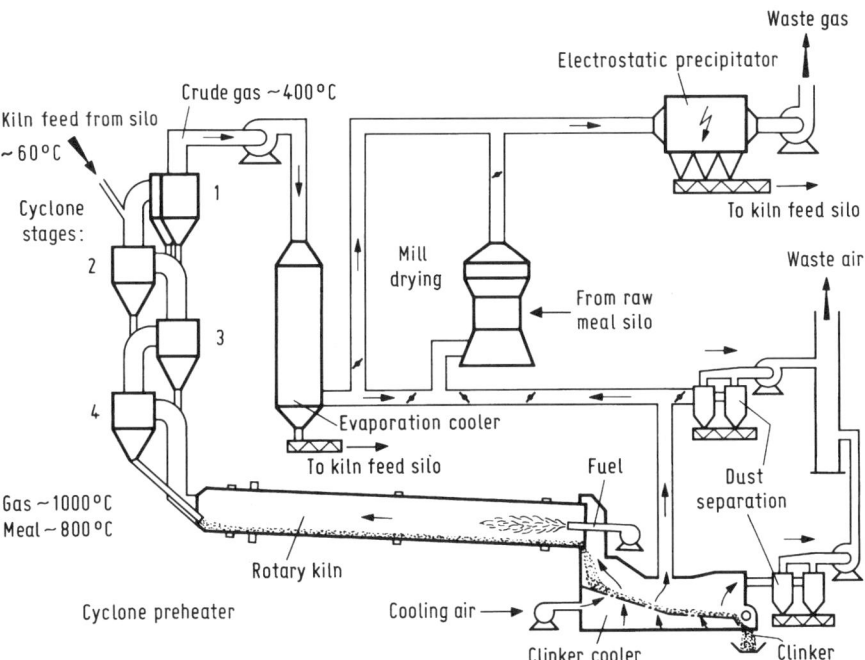

Fig. 9.2. Schematic presentation of the clinker burning process with a cyclone preheater

directly to the electrostatic precipitator. There the dust is precipitated and then transported to the kiln feed silo. With the evaporation cooler in an off-mode, the crude gas is passed through the mill, thus drying the raw material while it is being milled. The dust from the mill, which is made up of dust from both the crude gas and the raw materials, is also precipitated in the electrostatic precipitator and fed to the kiln feed silo. The dust content of the crude gas leaving the cyclone preheater is about 50 to 150 g/m^3. The dust content of the waste gas emitted is reduced by the electrostatic precipitator to less than 50 mg/m^3.

9.2.4
Sampling Points

Most of the sampling is done at points where material is being transferred from one stage of production to the next. For example, the material is transported by a conveyor belt. At the end of this it drops onto another belt or into a silo. From the silos it is drawn off by conveyor belts. The kiln feed is separated in the cyclones. Thereafter it falls through pipes into the gas ducts. Samples may be drawn from these pipes or from the input or output pipes of the pneumatic conveyors. When choosing sampling points, it should be remembered that all the different materials

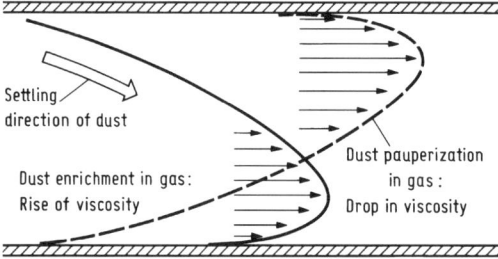

Fig. 9.3. Settling behaviour of dust in a horizontal pipe

are distributed according to particle size. When these materials are slightly shaken for example, they move and tend to segregate, so that the larger particles migrate to the edge and the smaler ones to the center.

For sampling from dust-bearing gases, a place should be chosen where laminar upward flow exists since from any other direction a representative sample cannot be gained. This is evident from Fig. 9.3, in which the conditions in a horizontally flowing gas are illustrated.

In a flowing gas, dust tends to settle. This leads to an enrichment of dust in the lower part and a depletion in the upper part of a duct. Therefore, the gas will flow faster in the upper, impoverished, part of the pipe than in the lower, enriched, one. Some particular sampling points will be discussed in Chap. 9.3.

9.2.5
Determining Masses

To calculate the mass balance of an element, the masses of all flowing materials have to be determined. The raw materials and sometimes the kiln feed and the coal pass over conveyor balances or weighing bins [20, 21]. Conveyor balances indicate the mass flow by the lateral deflection of measuring rollers and the speed of the conveyor belt. The stored weight of a silo is normally determined by load cells, sometimes by the filling position. To measure the flowing mass, the discharge time under normal conditions must be known. The amount of clinker produced is determined by directing the total production for a certain period, for example half an hour, to a particular bin. Afterwards the contents of the bin are transferred to trucks and these are weighed. When calculating the clinker production, the fuel consumed by the trucks has to be considered. The fuel oil is measured using calibrated volume meters. From this, the fired mass stream is calculated with the density of the oil under operating conditions.

9.2.6
Frequency of Sampling

Although the quality of the clinker produced very much depends on the homogeneity of the raw materials, there are minor changes in the composition of the materials involved. Therefore, repetitive measurements of every single cement

Sampling of Industrial Material

clinker burning process are necessary before an assessment of the long-term behaviour of a particular element can be made. With a steadily working kiln it is normally sufficient to repeat the sampling every hour over a period of twelve hours [8, 18, 22]. In some difficult cases all sampling procedures have to be performed every half an hour or the sampling and measurements have to be carried out at intervals of one hour over a period of 24 hours [23].

9.3
Sampling

The sampling procedure has to be adapted to the material and to the equipment used for transportation. Furthermore most materials have a range of particle sizes and therefore tend to segregate.

9.3.1
Conveyor Belts

The raw materials, raw meal and sometimes milled coal are transported by conveyor belts [24–26], an example of which is shown in Fig. 9.4. The transported material falls from the discharge end pully into a bin or onto another conveyor belt. The falling material can be collected in a box whose width must be greater than that of the material stream. At least 10 kg should be taken with each sampling step.

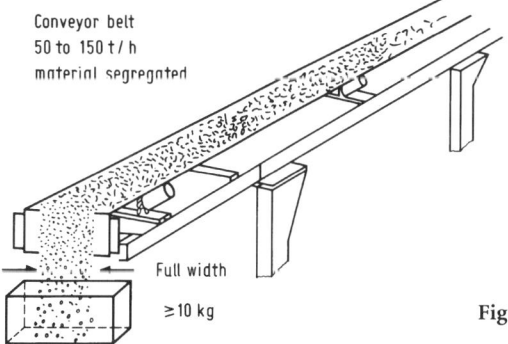

Fig. 9.4. Sampling from a conveyor belt

9.3.2
Pneumatic Conveyors

Powdery materials such as kiln feed or milled coal are often transported in pneumatic conveyors, two types of which are shown in Fig. 9.5. They are normally inclined at an angle of between 4 and 8° [27].

In the upper part of the figure the material flows on a steady air stream, which passes a porous plate. The lower part shows a conveyor that is mainly used for self-

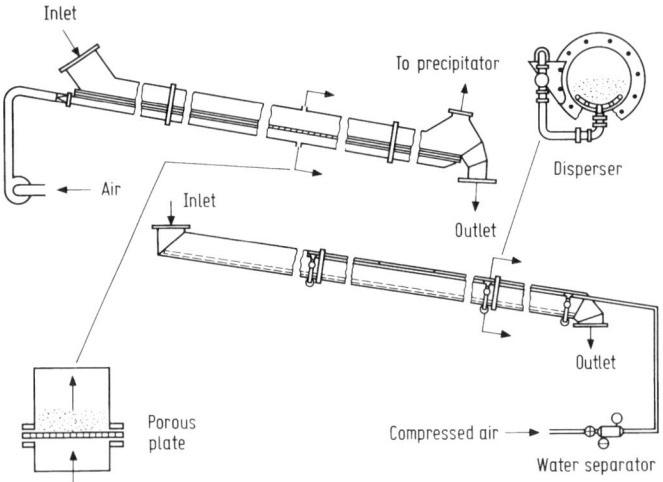

Fig. 9.5. Schematic presentation of aerated pneumatic conveyors

flowing materials, which have to be dispersed from time to time by a pressure shock. The gas phase above the milled coals has to be inert. Regardless of the material being transported the pneumatic conveyors have to be airtight. At the end of the conveyor the material falls into a chute, into which a charging valve might be built. Figure 9.6 shows such a valve, into which a double-pipe with slits may be pushed until its end is flush with the back wall of the chute [28]. While introducing the double-pipe the slits are kept closed, as can be seen in the lower part of the figure. For sampling the inner pipe is turned. Now both slits are open towards the falling material. For sampling the double-pipe is closed by turning the inner pipe and drawing it out of the valve, which has to be closed simultaneously. The pipes are emptied into a suitable vessel. The weight of the sample should be at least 500 g; otherwise the sampling will have to be repeated.

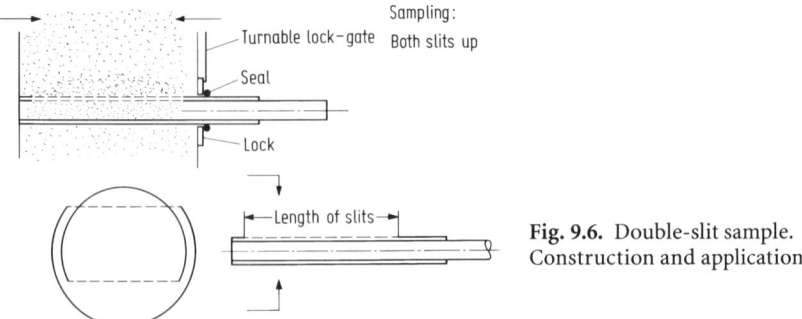

Fig. 9.6. Double-slit sample. Construction and application

9.3.3
Used Tires

Several cement plants use tires as an alternative for fuel. The tires mostly run through a groove to a balance, with which the weight of each tire is determined. Afterwards the tire drops through a sluiceway into the kiln. From the row of tires in front of the balance a single tire is taken and weighed. From the sampled tire a strip of about 2 cm width is cut, for example using an abrasive disc [4, 5]. The tire has to be cut at right angles to the circumference, as can be seen in Fig. 9.7, because different kinds of rubber are used for the bead, the carcass, the belt and the tread [29]. Therefore, only by cutting the tire in this way can a representative sample be obtained.

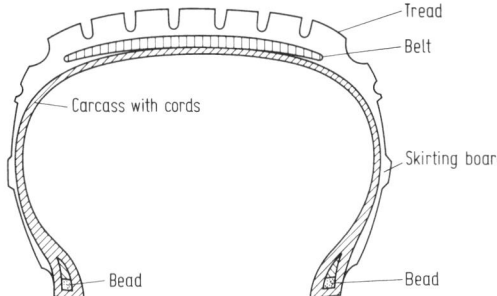

Fig. 9.7. Cross-section of a strip of tire, as used for sampling

9.3.4
Fuel Oil

The share of the Fuel oil represents only about 5% of the total fuel consumption. Since its viscosity is very high pumping the fuel oil into the burner nozzle and igniting it is only possible at elevated temperatures. It is therefore heated up in high pressure containers to about 110 to 120°C and pumped at a pressure of approximately 20 bar into the burner. Between the pipeline and the burner a sample of about 1 l is taken. Simultaneously the temperature is determined. In the heating and pumping stages the consumed fuel oil is measured by calibrated volume flow meters [27]. The mass flow rate of the fuel oil can be calculated from this volume and its density, which has to be determined in the laboratory at the operating temperature [23].

9.3.5
Crude-Gas Dust

From the crude gas/dust mixture a sampling stream with a flow rate of about 6 m³/h is drawn off isokinetically and sampled in the dust bag of the apparatus shown in Fig. 9.8 [30]. Isokinetic sampling means that the sampled gas stream is moving at the same velocity as the gas from which the sample has been drawn.

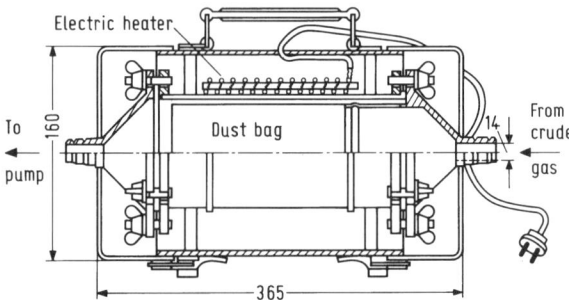

Fig. 9.8. Device for the sampling of dust in crude gas

The apparatus consists of a metallic housing, which is thermo-insulated and maintained by an electric heater at a temperature above the dew point; this constant temperature will avoid condensation of water vapor during sampling. Since the crude gas contains about 50–150 g dust/m^3 the dust bag can only operate for between 3 and 6 minutes before the pores of the filter are blocked and the dust bag has to be exchanged. With this apparatus a sample of about 30–60 g can be taken at any one time.

9.3.6
Precipitated Dust

Because electrostatic precipitators are mostly operated at a slightly reduced pressure, direct sampling and determination of the flowing masses is not possible. Instead of direct sampling, samples of the crude gas are taken in front of the precipitator (Sect. 9.3.5) and samples of the filtered gas (Sect. 9.3.7) behind it. Additionally in both cases, the dust contents of the gases have to be determined.

9.3.7
Waste-Gas Dust

The dust content of the emitted waste gas should not exceed 50 mg/m^3. Figure 9.9 shows a standardized device for the sampling of dust in waste gases [31, 32]. The sampled gas stream is drawn into a silicia cup where it is absorbed into 2 g of compacted silica wool. From the increase in weight of the silica wool and the volume of the sampled gas the dust content can be calculated. With this method the dust in a sampling stream moving at about 5 m^3/h may be taken for about 30 min, during which time a dust mass of between 10 and 30 mg will be obtained. To determine the metal content, the dust has to be decomposed together with the silica wool [33].

Sampling of Industrial Material

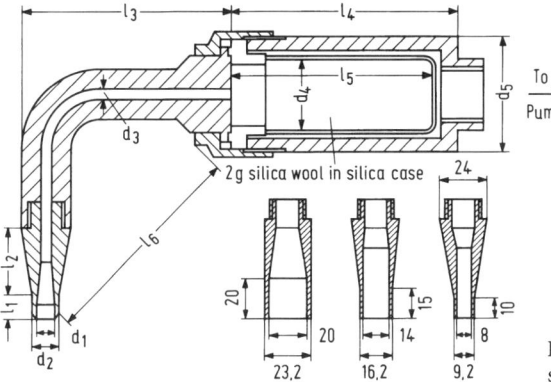

Fig. 9.9. Device for the isokinetic sampling of dust in waste gas

9.3.8
Volatile Elements

According to more recent ideas, emissions that pass through the dust filters should also be determined [34, 35]. The method used for this determination has been repeatedly discussed in the literature [36–38], particularly in the case of mercury emissions.

In Fig. 9.10 an apparatus is shown for the determination of volatile elements in the waste gas [39]. On the left, a container filled with silica wool is used for separating off the dust. In order to minimize the intake of dust, the suction holes are located on the opposite side of the container to those through which the waste gas enters. The waste gas is diverted through the silica wool and its velocity is decreased. In this way the effectiveness of the wool to absorb dust is maxmized. As a rule, the gas will now be practically free of dust but if the silica wool has not been entirely effective, any remaining dust will be removed when the gas is passed through the additional filter shown in Fig. 9.10 [40]. All parts of the apparatus situated outside of the chimney, and up to the junction where the gas-washing bottles (not shown in the figure) are attached, are maintained at a high enough temperature that will guarantee that volatile substances will not condense on the inside of the apparatus.

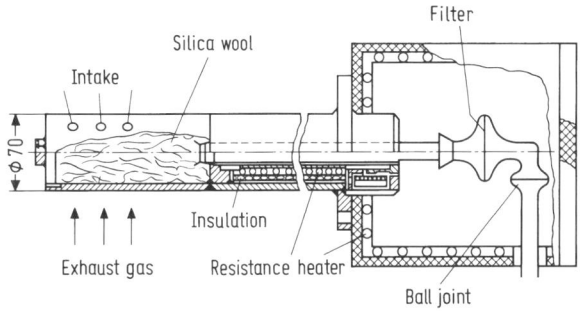

Fig. 9.10. Device for sampling volatile constituents in waste gas

9.4
Sample Preparation, Decomposition and Analysis

The samples taken in the course of a balancing experiment have to be prepared for the chemical analysis [33]. To this end, they are crushed, separated and, if necessary, milled. Weighed portions of these samples are combined to make a total sample, in which the weight of each portion corresponds to the material flow at the time when the sample was taken [18, 22]. This produces a mixture which corresponds to a weighted-mean mixture of the material from one sampling point during the time of the experiment. This representative sample is decomposed (Chap. 14) and analysed [33, 41–43]. The dust samples from the waste gas (Sect. 9) cannot be combined in this way and therefore have to be digested and assessed separately.

9.5
Checking Mass Flows

The results of the chemical analyses (Sect. 9.4) and the measurements of the mass-flow rates (Sect. 9.2.5) are used to balance some of the non-volatile major and minor elements [22]. In the case of the clinker burning process these elements are calcium, silicon, iron and aluminium. As an example, the balances of calcium and aluminium are given in Table 9.2.

In the first column the materials and their mass-flows rates are given. The middle column contains the balance for the major element calcium. The balance deficit is only 316 kg/h (or 0.5% of the balance sum) and lies on the input side of the balance. The right-hand column shows the balance for the minor element aluminium. The balance deficit is also low (0.6%) but lies on the output side of the balance.

Both balance deficits are smaller than 2% and are both on the input and output sides of the balance. They must therefore be the result of inevitable errors in mass-flow measurements, sampling, sample preparation, solubilization and analysis. From this result it can be concluded that the mass-flow rates of the materials were

Table 9.2. Mass flow of materials and balance of calcium and aluminium [22]

Material mass flow in kg/h		Calcium mass flor		Aluminium mass flow	
		in kg Ca/h	in %	in kg Al/h	in %
Kiln feed	210400	64990	99.0	4142	91.3
Fuels	15900	358	0.5	394	8.7
Balance deficit		316	0.5		
Input		65664	100.0	4536	100.0
Clinker	128000	60563	92.2	4173	92.0
Crude gas dust	16100	5096	7.8	336	7.4
Waste gas dust	11	5	0.0_0	0.5	0.0_1
Balance deficit				26.5	0.6
Output		65664	100.0	4536	100.0

Table 9.3. Mass flow of materials and balance of vanadium and zinc [22]

Material mass flow in kg/h		Vanadium mass flow		Zinc mass flor	
		in kg V/h	in %	in kg Zn/h	in %
Kiln feed	210 400	7.869	95.3	13.255	97.4
Fuels	15 900	0.318	3.8	0.358	2.6
Balance deficit		0.074	0.9		
Input		8.261	100.0	13.613	100.0
Clinker	128 000	7.589	91.9	11.520	84.6
Crude gas dust	16 100	0.671	8.1	1.143	8.4
Waste gas dust	11	0.001	0.0_1	0.001	0.0_0
Balance deficit				0.949	7.0
Output		8.261	100.0	13.613	100.0

measured with sufficient accuracy and can therefore be used for balancing the mass-flow rates of trace elements. This statement is the basis of the following balances for non-volatile trace elements.

Table 9.3 gives the balances of the trace elements vanadium and zinc. The first column contains the same materials and mass flows as in Table 9.2. The middle column for vanadium shows that the balance deficit is only 0.9 % lies on the input side of the balance. A similar result is given for zinc in the right-hand column but here the balance deficit is on the output side and amounts to 7 % only. The deficits of both elements are smaller than 10 % and are to be found on both the input and output sides of the balance. Therefore, they can be neglected compared to the balance sum.

The balances for the trace elements arsenic, beryllium, lead, cadmium, nickel, vanadium and zinc, taken from a large number of clinker burning processes, showhed that the balance deficit was always small and always distributed statistically to the input and output sides of the balance [22]. In those cases where the balance deficit exceeded 10 %, it could be shown that errors had occured in sampling, sample preparation or in the analyses. This finding shows that these non-volatile trace elements behave in a similar way to the major or minor elements previously mentioned. In the case of other trace elements, thallium in particular, there were always systematic balance deficits on the output side of the balance, although the major and minor elements showed no significant balance deficit. This behaviour indicates that enrichment by vaporization had taken place during the whole process.

9.6
Conclusions

The following conclusions can be drawn:

1. All samples of all balance mass streams have to be analysed. The sum of each complete analysis has to be 100.0 ± 0.2 % by weight.

2. When non-volatile elements are balanced the deficit should not exceed (a) 2% of the balance sum for the major and minor elements and (b) 10% of the balance sum for trace elements.
3. The balance deficit should be statistically distributed between the input and output sides of the balance for all elements. If this is the case, the balance can be regarded as correct. If the deficit systematically lies on either the input or the output sides of the balance, then either the analysis is at fault or the element in question is volatile. For an assessment of the behaviour of volatile elements further investigations will be needed.

9.7
References

1. Goes C (1960) Über das Verhalten der Alkalien beim Zementbrennen. Schriftenreihe der Zementindustrie, Heft 24, Verein Deutscher Zementwerke, Düsseldorf, Thesis
2. Sprung S (1964) Das Verhalten des Schwefels beim Brennen von Zementklinker. Schriftenreihe der Zementindustrie, Heft 31, Beton-Verlag, Düsseldorf, Thesis
3. Sprung S, Seebach HM von (1968) Zement-Kalk-Gips 21:1
4. Sprung S, Rechenberg W (1978) Zement-Kalk-Gips 31:327
5. Sprung S, Rechenberg W (1973) Zement-Kalk-Gips 36:539
6. Sprung S, Kirchner G, Rechenberg W (1984) Zement-Kalk-Gips 37:513
7. Kirchner G (1985) Zement-Kalk-Gips 38:535
8. Kirchner G (1986) Das Verhalten des Thalliums beim Brennen von Zementklinker. Schriftenreihe der Zementindustrie, Heft 47, Beton-Verlag, Düsseldorf, Thesis
9. Billings CE, Matson WR (1972) Mercury Emissions from Coal Combustion. Science 176:1232
10. Billings CE, Sacco AM, Matson WR, Griffin RM, Coniglio WR, Harly RA (1973) J Air Poll Contr Assoc 23:773
11. Kaakinen JW (1974) Trace Element Study in a Pulverized-Coal-Fired Power Plant. Thesis, University of Colorado
12. Kalb GW (1975) Amer Chem Soc, Advanc Chem Series 141:154
13. Klein DH, Andren AW (1975) Water Air Soil Poll:71
14. Heinrichs H (1977) Naturwissenschaften 64:479
15. Anderson WL, Smith KE (1977) Environm Sci Technolog 11:75
16. Kautz K, Kirsch H, Laufhütte DW (1975) VGB Kraftwerkstechn 55:672
17. Wischers G (1988) Möglichkeiten und Grenzen des Recyclings in der Zementindustrie. Vorträge auf der Herbsttagung 1988 des Wissenschaftlichen Rates der AIF über Umweltfreundlichkeit industrieller Produktionsprozesse, Köln: Arbeitsgemeinschaft industrieller Forschungsvereinigungen (AIF) 1
18. Sprung S (1982) Technologische Probleme beim Brennen des Zementklinkers, Ursache und Lösung. Schriftenreihe der Zementindustrie, Heft 43, Beton-Verlag, Düsseldorf. Habilitation. 2nd, engl Ed (1985) Technological problems in pyroprocessing cement clinker: Cause and solution. Beton-Verlag, Düsseldorf
19. VDI Guideline 2094 (1985) Emission Control Cement works. (German). Düsseldorf
20. Schlepütz H (1976) Aufber Techn 17:11
21. Merks JW (1985) Sampling and Weighing of Bulk Solids. Trans Techn Publications, Clausthal-Zellerfeld
22. Kirchner G, Rechenberg W (1986) Spurenelementbilanzen von Zementöfen. In: Welz B (ed) Fortschritte in der atomspektrometrischen Spurenanalytik, Vol 2, VCH Verlagsgesellschaft Weinheim
23. VDZ Working Group Kiln Experiments (1992) Executing and Assessing Kiln Tests. Guideline Vt 10. Verein Deutscher Zementwerke, Düsseldorf
24. Taubmann HJ (1984) Aufber Techn 25:189

25. Karalus E (1981) TIZ-Fachber 105:792
26. Sporbeck H (1965) Z Anal Chem 209:60
27. Duda WH (1977) Cement Data Book, 2. Ed, Bauverlag Wiesbaden Berlin
28. Alberti K, Mälzig G (1962) Zement-Kalk-Gips 15:262
29. Schneider P, Dane J, Ecker R, Hühnen W, Oettner K, Reissinger S, Sim G, Spulak F v (1957) Kautschuk. In: Foerst W (ed) Ullmanns Encyklopädie der technischen Chemie. Urban und Schwarzenbach, München Berlin, Vol 9, 3 Ed, p 305
30. Verein Deutscher Zementwerke (1962) Technique of Measurements in the Cementindustriy (German) 07.23, Düsseldorf, p 1
31. VDI Guideline 2066, Part 1 (10.75) Particulate Matter Measurement – Over-view. Commission on Environmental Protection. Beuth, Berlin Köln
32. VDI Guideline 2066, Part 2 (Draft 01.89) Particulate Matter Measurement – Measuring Dust in Streaming Gases. Gravimetric Determination. Beuth, Berlin Köln
33. Working Group Analytical Chemistry (1993) Determination of Trace Elements in Materials of the Cement Process (German). Beton-Verlag, Düsseldorf
34. Jockel W, Hönig H-J, Mistele J (1986) Standardisierung der Emissionsmessung toxischer Staubinhaltsstoffe. Umweltbundesamt, Berlin, Forschungsbericht 86–104 02 157 sowie supplement (1987), addenda 1988
35. VDI Guideline 3868, Part 1 (Draft 10.92) Determination of Total Emission of Metals, Metalloids, and their Compounds. Manual Measurement in Flowing, Emitted Gases. Sampling System for Particulate and Filter-Passing Matter.
36. Dannecker W, Redmann WA, Düwel U (1985) Staub, Reinhalt Luft 45:331
37. Redmann WA (1986) Entwicklung neuartiger Probenahmesysteme zur Erfassung filtergängiger Metalle und Metalloide sowie deren Verbindungen aus Rauchgasen. Thesis, University of Hamburg
38. Bachmann G, Rechenberg W (1989) Aufschluß von Silikaten für die atomspektrometrische Quecksilberbestimmung. In: Welz B (ed) 5. Colloquium atomspektrometrische Spurenanalytik, Bodenseewerk Perkin-Elmer
39. Kuhlmann K, Kirchartz B, Rechenberg W, Bachmann G (1991) Zement-Kalk-Gips 44:209
40. Haegermann B (1982) Dampfförmige Schwermetallverbindungen im Zementofenabgas. (MS Work), Technical University Clausthal
41. Ludwig R, Richartz W (1978) Zement-Kalk-Gips 31:550
42. VDZ Working Group XRF (1978) Zement-Kalk-Gips 31:558
43. VDZ Working Group Analytical Chemistry (1970) Schriftenreihe der Zementindustrie, Vol. 37 (German). Beton-Verlag, Düsseldorf

Chapter 10

Food Products of Animal and Plant Origin

Lothar Matter and Markus Stoeppler

10.1 Introduction

Trace elements occur in biological systems only in relatively small amounts, i.e. in the order of mg/kg or below. They have been part of the natural environment throughout historical times. Although trace elements such as selenium, iodine, molybdenum and cobalt are generally known to be physiologically essential they might have adverse effects on human health if larger amounts are ingested. Thus, beneficial or harmful effects are basically dose-dependent.

On the other hand, elements which are today considered to be toxic to man and animals, such as lead, cadmium and mercury, might play an essential, as yet

Table 10.1. Typical cadmium levels in a variety of foodstuffs from several countries [2]

Cd content [µg/kg fresh weight]	Type of food	Remarks
> 200	Some mushrooms, some cocoa powders, dark chocolate, blue poppy seeds, semi-finished products from poppy seeds, marine mussels	Mushrooms differ strongly in Cd contents; dark chocolate from particular cocoa species represent only a small percentage of total chocolate products
≤ 200	Kidneys and livers from pigs, cattle and sheep, most mushroom species, some baked goods with blue poppy seeds	
≤ 40	Wheat, wheat flour, wheat bread, bran, potatoes, root and foliage vegetables, rice, shrimps	Cadmium content may vary significantly in rice
≤ 20	Rye flour, rye bread, beans, tomatoes, fruit, eggs, fresh water fish	Fruit and fresh water fish frequently have Cd levels up to 10 µg/kg
≤ 5	Meat from poultry, pigs, cattle and sheep, fillet of marine fish, wine, beer, fruit juices	
≤ 1	Tap water, milk products	Studies performed during the last decade have shown that Cd in dairy milk on average is < 0.1 µg/kg

Table 10.2. Mercury levels in different foods [excluding fish that may contain much higher (mg/kg) Hg levels, mainly in the form of methyl mercury], Federal Republic of Germany [3]

Food type	Average or range [µ/kg actual weight]
Fresh milk, yoghurt, white cheese	≤ 0.02 – 0.04
Other cheese	≤ 0.05 – 0.65
Condensed milk	≤ 0.04 – 0.15
Butter	≤ 0.02
Margarine	≤ 0.02
Whole eggs	0.2 – 3.0
Meat (pork, beef, fillet)	1.7 – 3.3
Chicken meat	0.2 – 0.4
Ham and sausages	0.24 – 4.7
Kidney (pig, beef)	6.0 – 12
Liver (pig, beef)	2.2 – 3.0
Potatoes	≤ 0.02 – 0.08
Vegetables	≤ 0.02 – 1.55
Fruit	≤ 0.02 – 0.6
Wheat flour	0.3 – 0.5
Bread (various)	0.7 – 2.8
Noodles	≤ 0.2 – 0.5
Bran	1.25 – 16.4
Rice	0.6 – 5.0

unknown, role, possibly at concentration levels that still cannot be determined reliably or even not at all [1].

The absorption of trace elements in the gastro-intestinal tract and their effect on the human organism are predominantly determined by their chemical form ("speciation") (e.g. inorganic or organic mercury compounds), the oxidation state [e.g. As(III) versus As(IV), Cr(III) versus Cr(VI)] and the composition of the food consumed.

"Normal", i.e. nonessential, trace elements are usually not intentionally added to food products. They enter the environment either from geological or, and often significantly, from man-made sources and thus occur partly in the food chain to man and animals. Examples of typical levels of cadmium and mercury in basic food and food products from recent studies are presented in Tables 10.1 and 10.2 [2, 3].

10.2
Recommendations, Standards and Estimations of the Tolerable Intake of Toxic Metals via Food

In 1972 the Joint WHO/FAO Expert Committee on Food Additives, following the evaluation of certain food contaminants and additives, recommended provisional tolerable weekly intakes for humans of mercury, methyl mercury, lead and cadmium (Table 10.3 [4]), and somewhat later tentatively fixed the "maximum acceptable load of arsenic" as 0.05 mg per kg body weight per day [5]. Ewers, however,

Table 10.3. Provisional tolerable weekly intake of mercury, lead and cadmium

Element/compound	Provisional tolerable weekly intake for man		Acceptable daily intake for man
	[mg/person]	[mg/kg body weight]	[mg/kg body weight]
Mercury			
– total mercury	0.3	0.005	None
– methyl mercury expressed as mercury	0.2	0.0033	None
Lead	3	0.05	None
Cadmium	0.4 – 0.5	0.0067 – 0.0083	

stated that, as far as these data are concerned, "these recommendations lack a sound toxicological basis" [6].

Subsequently, the maximum admissible levels of metals and metalloids for certain food products were issued in several countries. Here are some examples from [6]:

- *Italy*: Maximum admissible levels of metals in wine: lead 0.3 mg/l; copper, 1.0 mg/l; zinc, 5 mg/l; mercury in fish and shellfish, 0.7 mg/kg
- *Switzerland*: Mercury in fish and shellfish, 0.5 mg/kg. The Swiss Ordinance on Additives and Contaminants in Dietary Products, issued 1986, contains the maximum tolerable levels for aluminium, arsenic, cadmium, chromium, copper, lead, mercury, nickel, selenium, tin and zinc in beverages, fruit and vegetable.
- *Germany*: The Wine Ordinance gives the maximum admissible concentrations in wine and liqueurs for aluminium, 8 mg/l; arsenic, 0.1 mg/l; cadmium 0.01 mg/l; copper, 5 mg/l; lead, 0.3 mg/l; tin, 1 mg/l; and zinc, 5 mg/l. For mineral and table waters the maximum admissible contaminant levels are the same as set by a CEC Directive for drinking water [7].

Since 1979 the German Federal Health Agency has been issuing guideline values for cadmium, lead and mercury in and on foodstuffs. The current, updated version of these guidelines, including thallium in fruit and vegetables, issued 1995 [8] is given in Table 10.4. These values represent the upper normal limits of the above metals presently available in dietary products in Germany. They should be used as guidelines for the responsible authorities and the food industry.

The investigation of guideline and limit values, that are usually at the µg/kg level, is very demanding for the analyst. To obtain accurate data he has to obey strictly the rules of trace analysis. Besides up-to-date instrumental equipment, tailored to the problems to be tackled trace analysis requires extensive experience with the initial steps (sample collection and sample preparation) that are essential for analytical reliability. These are commonly the parts of an analytical procedure that are prone to the largest irreproducible errors [9].

Food Products of Animal and Plant Origin

Table 10.4. Guideline values for cadmium, lead, mercury and thallium in and on foodstuffs, recommended for 1995/96 by the German Federal Health Agency. Units: mg/kg fresh weight or mg/l relating to the edible constituents, for dried products related to the re-hydrated product [8]

Food type	Lead	Cadmium	Mercury
Milk	0.03	0.005	0.01
Condensed milk	0.30	0.05	0.01
Cheese (except hard cheese)	0.25	0.05	0.01
Hard Cheese	0.50	0.05	0.01
Eggs	0.25	0.05	0.03
Beef, veal, pork, ground and chicken meat	0.25	0.10	0.03
Liver (beef, calf, pork)	0.50	0.30	0.10
Kidney (beef, calf, pork)	0.50	0.50	0.10
Meat products, sausages	0.25	0.10	0.05
Fish, fish products	0.50	0.10	0.50[a] Hg 1.0[b]
Crust-, shell- and soft animals except crustaceae and mussels	0.50	0.50	0.50[a]
Crustaceae	0.50	0.10	0.50[a]
Lobster	0.50	0.50	0.50[a]
Mussels	0.80	0.50	0.50[a]
Wheat grains	0.30	0.10	0.03
Rye grains, rice grains	0.40	0.10	0.03
Poppy seeds		0.30	
Linseed		0.60	
Sunflower seeds		0.60	
Sesame		0.80	
Nuts	0.50	0.05	0.03
Peanuts[c]	0.50	0.10	0.03
Peanuts, roasted[c]	0.50	0.10	0.03
Potatoes	0.25	0.10	0.02
Green vegetables	0.80	0.10	0.05
Spinach	0.80	0.50	0.05
Kitchen herbs, parsley leaves	2.00	0.10	0.05
Sprout vegetables	0.50	0.10	0.05
Fruit and root vegetables	0.25	0.10	0.05
except tuberous celeriac	0.25	0.20	0.05
Berries, pomaceous fruits, fruits with stones, citrus fruit, other fruit, parts of exotic vegetables, rhubarb	0.50	0.05	0.03
Refreshing drinks	0.20	0.05	0.01
Wine	0.30[a]	0.01[a]	0.01
Beer	0.20	0.03	0.01
Chocolate		0.40	
Milk chocolate and chocolate candies		0.15	
Cream chocolate		0.40	

[a] Regulation values 1995 for mercury in fish and fish products and shell animals as well as for lead and cadmium in wine.
[b] mg/kg for several fish species and fish products from these species: eel, pike, salmon, pike-perch, blue ling, sleeper shark, parbeagle, catfish, rosefish, swordfish, sturgeon, white halibut.
[c] Without brown seed skin.
Regulation values 1995/96 for thallium in fruits and vegetables: 0.1 mg/kg (fresh weight).

10.3
Sample Collection

Collection of food samples is usually performed randomly. The persons authorized for food inspection (e. g. food controllers or police officers) are entitled to collect by buying samples of all products that are covered by the Food and Consumption Articles Act (Sect. 42 LMBG [10]). Usually two samples of the same material, of the same lot and of the same manufacturer are taken. The sampling procedure depends on the kind of sample (tin, foil package, etc.). One of these samples constitutes the "official", the other the control sample. In cases of doubt the manufacturer or importer is thus able to use the officially sealed control sample and call upon an independent food expert to provide a counter opinion.

Only in cases where it is suspected that upper limits have been exceeded or guidelines not adhered to does a representative amount of sample have to be taken in accordance with the Food and Consumption Articles Act (Sect. 35 LMBG), provided the producer and the control laboratory are located at the same place. An example is sampling for the determination of pesticides on and in fruit and vegetables. In lots of 50 kg three, in lots of 50 – 100 kg at least five, and in losts of < 500 kg at least ten single samples have to be taken, if possible from different positions distributed over the whole plot. If the goods to be controlled have rotted, no samples should be collected.

If the weight of a lot is unknown or cannot be estimated satisfactorily, or if the goods are frozen, the following procedure has to be followed: out of 1 – 25 packages at least one, out of 26 – 100 at least five and out of > 100 packages at least ten samples have to be taken. By combining and mixing the single samples a composite sample has to be prepared, which then constitutes the final sample. Appropriate reduction procedures (see also Chapter 6) will finally lead to the laboratory (analytical) sample.

10.4
Sample Preparation

For the determination of trace elements in and on food products, e. g. in Germany, the prescriptions of Sect. 35 LMBGX-1 [11] have to be applied. Generally, only the edible constituents in the commercial form have to be analyzed. Certain parts (e. g. outside and wrapping leaves, shells, bones, connective tissue and tendons) have to be discarded. Significant dirtiness, e. g. adhering soil, rotted parts of leaves and plants, have to be discarded. For vegetables with a large surface area, e. g. green kale, washing is mandatory. Whether for this purpose water of defined purity or simply tap water has to be applied depends on the analytical task (which element, what concentration level).

10.5
Homogenization

Homogenization is in most cases an important step in the analysis of food and food products. The required degree of homogeneity depends predominantly on the size of the subsample taken for analysis. Inhomogeneities should be identified by parallel analysis of several subsamples of the same homogenate. Regrettably, in the current analytical routine often only one single subsample is analyzed, which can be a major source of error.

Voluminous and tough material should be pre-crushed with appropriate tools such as knives, mixers and mills. Firm materials such as cereals can be crushed by treating with hammer or table mills and thus simultaneously homogenized. Soft material should be homogenized with mixers, preferably with high-speed rod mixers. Another possibility in this case is freeze drying followed by pulverization. If the material is fibrous or relatively dry, homogenization can be facilitated by adding a defined amount of water of suitable purity.

Since homogenization procedures may introduce noticeable amounts of the elements to be determined, it is necessary to minimize, or if possible to avoid, these contamination effects. Tools for homogenization consisting of stainless (Cr/Ni) steel can be only used if there is proof that use of such a tool does not generate a detectable increase of chromium or nickel in the analyzed material. A titanium coating on crushing knives, for example, can avoid such a contamination. For titanium determination in food, of course, this tool is not suitable.

Many types of mills are equipped with balls of various materials that often show abrasion and thus can also contaminate the final homogenized sample.

A temperature increase during homogenization is also undesirable, because it may lead to losses if relatively volatile elements have to be determined. Therefore, heating effects during homogenization should be strictly avoided in such materials.

As a general rule it must be stated that contamination is never reproducible!

10.6
Decomposition Procedures

Due to the frequently very low contents of trace elements at the µg/kg level (fresh weight, see Tables 10.1 and 10.2) in food, the decomposition step is of particular importance. Thus, the applied decomposition procedures have to be properly selected as far as blank values and completeness of decomposition are concerned. From this it is obvious that decomposition procedures have to be carefully adapted to the particular task, i.e. the material to be analyzed, the elements and, very important, the elemental levels in this material.

The fundamental requirements to achieve reliable decomposition procedures can thus be stated as follows:

- the reagents used should be of the utmost purity to guarantee low blanks, which requires the use of e.g. acids purified by subboiling distillation (see Chapter 11);
- the material of the decomposition vessels should be as pure and chemically inert as possible and should possess a very low adsorption behaviour for trace elements (ultrapure quartz, very pure and dense PTFE);
- the ratio of the surface of the decomposition vessel to sample weight should be as small as possible in order to minimize errors due to adsorption and desorption of elements;
- the decomposition system should be closed to avoid volatilization losses.

Table 10.5. Trace-element content of various materials used for decomposition and/or storage of solutions from decomposition (values in µg/kg)

Element	PTFE	Quartz	Ultrapure quartz (Suprasil)	Laboratory glass (Borosilicate)
B		100	10	Main element
Na	25000	1000	10	Main element
Mg		10	100	6×10^5
Al		30000	100	Main element
Si		Main element	Main element	Main element
Ca		800–3000	100	10^6
Ti		800	100	3000
Cr	30	5	3	3000
Mn		10	10	6000
Fe	10	800	200	2×10^5
Co	2	1	1	100
Ni				2000
Cu	10	70	10	1000
Zn		50	200	3000
As		80	0.1	500–22000
Cd		10		1000
Sb	0.4	2	1	8000
Hg	10[a]	1	1	

[a] Depends very much on storage (working) conditions.

The following tables show typical trace element contents in various materials commonly used for the production of decomposition vessels (Table 10.5) as well as in ultrapure water and in acids with different degrees of purity (Table 10.6) [12].

Because of these particular requirements, decomposition in closed vessels (pressurized decomposition, see Chapter 12) has been implemented for the analysis of trace elements in food and food products. The official German method pursuant to Sect. 35 LMBG L 00.0019, "Pressurized decomposition, determination of lead and cadmium" in its first version was recently completely revised and extended. The

Table 10.6. Trace elements in suprapure water and acids of different purity (values in µg/l)

			Cd	Cu	Fe	Al	Pb	Mg	Zn
H₂O		subboiling	0.01	0.04	0.32	< 0.05	0.02	< 0.02	< 0.04
HCl	10 M	subboiling	0.01	0.07	0.6	0.07	< 0.05	0.2	0.2
HCl	10 M	suprapure	0.03	0.2	11	0.8	0.13	0.5	0.3
HCl	12 M	p. A.	0.1	0.1	100	10	0.5	14	8.0
HNO₃	15 M	subboiling	0.001	0.25	0.2	< 0.005	< 0.002	0.15	0.04
HNO₃	15 M	suprapure	0.06	3.0	14	10	0.7	1.5	5.0
HNO₃	15 M	p. A.	0.1	2.0	25	10	0.05	22	3.0
HF	54 %	subboiling	0.01	0.5	1.2	2.0	0.5	1.5	1.0
HF	40 %	suprapure	0.01	0.1	3.0	1.0	3.0	2.0	1.3
HF	54 %	p. A.	0.06	2.0	100	5.0	4.0	3.0	5.0

new version is called "Determination of trace elements in food. Part 1: Pressurized decomposition" [13]. This procedure describes the decomposition of previously homogenized samples either in closed PTFE vessels at 170 °C or quartz vessels at 320 °C (high pressure asher, HPA, see Chapters 12 and 14) with nitric acid purified by subboiling distillation. The decomposition with the HPA is necessary, if marine samples that contain the very stable compound arsenobetaine have to be analysed. The subsequent determination of iron requires the addition of a minimal amount of hydrochloric acid during decomposition in quartz vessels in order to avoid adsorption of iron traces onto the walls. For the subsequent determination of mercury, the use of quartz vessels is also prescribed. The sample weight has to be adjusted to the volume of the decomposition vessel and for safety reasons should never exceed the manufacturer's instructions.

The sample weight should also be selected by taking into account the carbon content of the analyzed material to achieve complete decomposition (see also Chapter 12). Because of an increased risk of contamination, the resulting solution from decomposition should in the case of extremely low metal contents, not be transferred into volumetric flasks. The use of graduated decomposition vessels or of inserts made of quartz for PTFE vessels is strongly recommended. These vessels allow decomposition and making up to volume in the same vessel. If there are not enough of these special decomposition systems available in the laboratory, it is possible to transfer the decomposed solution into flasks made of quartz glass (mandatory for subsequent mercury determination), FEP (fluoroethylpropylene) or PFA (perfluoralkoxy-polymers).

Other decomposition procedures such as wet ashing in open vessels or dry ashing in air with external heat sources (muffle furnaces, IR lamps) should never be used for the analysis of trace elements in food products. The same applies to perchloric acid, in the past frequently and still occasionally used for decomposition, because of its hazardousness and proven explosion risk if biological materials have to be decomposed.

10.7
Analysis and Quality Control

The analytical methods commonly applied for determining trace elements in food depend on the particular problem and the equipment of the laboratory. The currently applied methods are various modes of atomic absorption spectrometry, atomic emission spectrometry, mainly with plasma excitation (ICP-OES), voltammetry and, of increasing importance, plasma source mass spectrometry (ICP-MS).

To achieve relevant data, it is absolutely necessary to check the results by different methods, independently of each other, to participate in interlaboratory comparisons, and to use certified reference materials and/or reliably characterised reference samples.

The permanent use of certified reference materials (CRMs) in an analytical laboratory is mandatory [14]. This is a proven approach for identifying systematic analytical errors. For each laboratory, quality control with all the recommended procedures for that purpose should be given first priority [15]. Even a heavy routine workload should never prevent current direct or indirect quality control measures from being used. Control samples, which may consist of certified or other reference materials of appropriate homogeneity analysed in one's own laboratory, have to be continuously used to test the reliability of equipment and personnel. The control of blank values and their variation belong to the same, very important, category of mandatory quality control measures. It is of fundamental inportance to understand that simply the "number" of an analytical measurement cannot be taken on its own, but rather as a result of an analytical process and thus only in combination with methodological scatter, method used, detection and determination limits as well as numbers of independent determinations. The correct interpretation of all these terms for a relevant final value should thus be accepted and applied in general laboratory practice, as well as outside the rather narrow branch of trace-element analysis within analytical chemistry [9].

10.8
Consumption Recommendations

The consumer has somewhat limited means of protecting himself against unnecessary burdens of toxic heavy metals. An example of how a national authority can inform and help consumers is the "Consumption Recommendations" of the German Federal Health Agency, referred to below in some detail [16].

- *Fruit and vegetables should be washed and peeled.* Detailed investigations have proven that thorough cleaning or an appropriate preparation of vegetable foods can significantly reduce heavy metal pollution on the outer surface. Since one cannot expect that the consumer has detailed knowledge of whether the purchasedfoods originate from pollute areas, it is generally recommended that only thoroughly washed fruit and vegetables should be consumed. Particularly careful cleaning is necessary for parts of plants with rough or hairy surfaces (e.g. green kale, peaches). For fruit and vegetables that can be consumed washed or peeled, peeling is be preferred.
- *Offal from game should be avoided.* Particularly in the area of the former East Germany (GDR) high mercury contents have been observed, mainly for wild boar. This is due to the long-term application of mercury-containing disinfectants in agriculture.
- *Offal from older animals for slaughter should only be occasionally consumed.* Offal from cattle and pigs, especially from older animals, can frequently contain higher cadmium contents than other animal foodstuffs. It is thus recommended that kidneys and other offal should only be occasionally consumed, i.e. every two to three weeks.
- *Wild mushrooms should only be occasionally consumed.* The cadmium and mercury, and also radionuclide content (the latter deriving in Europe from the Chernobyl accident), of wild (but not of cultivated) mushrooms may be significantly higher compared to other vegetables. If wild mushrooms are regularly consumed, the weekly consumption should not exceed 200–250 g. But there is no basic objection to the occasional consumption of larger amounts of mushrooms. Children should, however, corresponding to their low body weight, con-

sume less wild mushrooms. The following species that frequently show especially high cadmium contents should be consumed not at all or not repeatedly: *Agaricus arvensis* and *Psalliopa arvensis* as well as the giant champignon species *Agaricus augustus* and *Agaricus perarus*.

During preparation of the above-mentioned mushrooms, the lamellae, the tube layer and, if possible, also the skin of the cap should be discarded, because the highest concentrations of heavy metals have been detected in these mushroom tissues.

If wild mushrooms are regularly consumed, the consumption of other food with high metal contents, particularly kidneys, should be avoided.

The German Federal Health Agency advocates that the consumer should follow these recommendations because the above-mentioned metal pollutants are generally undesirable, although in many cases unavoidable. Up to certain levels they may be quite safe, but they nevertheless contribute to the burden of heavy metals in the human body.

10.9 References

1. Schweizerisches Lebensmittelbuch (1989) Chapter "Trace Elements". Eidg Drucksachen und Materialzentrale, Bern
2. Stoeppler M (1991) Cadmium. In: Merian E (ed) Metals and Their Compounds in the Environment. VCH Publisher, Weinheim, p 803
3. May K, Stoeppler M (1983) Studies on the biogeochemical cycle of mercury. 1. Mercury in sea and inland waters and food products. In: Müller G (ed) Proc Int Conf Heavy Metals in the Environment. CEP Consultants Ltd, Edinburgh, Vol 1, p 241
4. WHO/FAO Joint Committee on Food Additives (1972) Evaluation of certain food additives and the contaminants mercury, lead and cadmium. WHO Tech Rep Ser 505
5. WHO (1973) Trace elements in human nutrition, WHO Tech Rep Ser 532
6. Ewers U (1991) Standards, guidelines and legislative regulations concerning metals and their compounds. In: Merian E (ed) Metals and Their Compounds in the Environment. VCH Publisher, Weinheim, p 687
7. Council Directive Relating to the Quality of Water Intended for Human Consumption (80/778/EEC) OJ No L of 30.08.1980, p 11
8. Bundesgesundheitsblatt (1995) 8/5:204–206
9. Pfannhauser W (1988) Essentielle Spurenelemente in der Nahrung. Springer, Berlin Heidelberg New York
10. Lebensmittel und Bedarfsmittelgegenständegesetz of 15.08.1974, version of 22.01.1991, BGBL I 121
11. Bundesgesundheitsblatt (1979) 22 No. 15
12. Tschöpel P, Kotz L, Veber M, Tölg G (1980) Fresenius Z Anal Chem 302:1–14
13. Amtliche Sammlung von Untersuchungsverfahren nach § 35 LMBG, Nr. L 00.0019 (1993). Beuth, Berlin
14. Parr R, Stoeppler M (1994) Reference materials for trace element analysis. In: Herber RFM, Stoeppler M (eds) Trace element analysis in biological specimens. Elsevier, Amsterdam, p 231
15. Griepink B, Stoeppler M (1992) Quality assurance and validation of results. In: Stoeppler M (ed) Hazardous metals in the environment. Elsevier, Amsterdam, p 517
16. BGA Pressedienst 01/1991

Chapter 11
Sample Preparation: an Introduction

MARKUS STOEPPLER

11.1
General Remarks

In the course of an analytical procedure the next step after sampling (and if necessary homogenization and reduction to obtain appropriate subsamples) is the preparation of the material for the determination of analytes (elements, organometallic compounds, anions etc.).

If the subsequent determination of organometallic compounds or the valency states of elements is required, after proper sampling (see Chapter 1) various extraction/separation steps are necessary. Although this book restricts itself to total element determination, because of the considerable and steadily increasing importance of speciation, the reader's attention should be drawn to several, mainly recent, reviews dealing with speciation and speciation methods in general in the environment [1–11], in biological samples [12, 13] and for anion determination [14, 15].

As far as the determination of the total elemental content alone is concerned, three general approaches have to be considered.

- Dissolution of the sample or analyte extraction in complex matrices; the latter can be facilitated with, for example, the aid of ultrasonification [16]. In both bases predominantly aqueous solutions result that usually can be directly analyzed by a variety of analytical methods (summarized in Table 11.1, after [17]). Another approach, the solubilization of relatively soft biological material such as human and animal brain and muscle tissue using quaternary ammonium hydroxides [18, 19], belongs to the same category. It was shown that such procedures are also useful for the determination of arsenic species [20].
- Partial or complete decomposition of an appropriate number of subsamples for determination of the analyte(s) (e.g. [21–25]). The methods applied have to be selected not only on the basis of the expected concentration of the analyte in the (diluted) solution from decomposition, but also according to general requirements such as mono-, oligo- or multi-element determination, numbers and mass of samples, and last but not least also according to the laboratory's equipment and the experience of the analytical staff. Table 11.1 presents a schematic overview of the methods currently used for this purpose and the quality of the decomposition procedure required for reliable results. Since some electrochemical methods are often vulnerable to incomplete decomposition, Chapter 14 treats this subject in some detail.
- Direct analysis of subsamples of the collected material. This is often feasible for aqueous samples after they have simply been acidified or biological fluids and

Table 11.1. Overview of the properties of the most frequently applied trace analytical methods [17]

Method	Potential	Dynamic range	Typical detection limit (analyte solution)	Completeness of decomposition required	Direct determination in solids and solutions
Photo	mono	+	mg/L	++	(yes)
F-AAS	mono	+	< mg/L	++	yes
H-AAS	mono	+	≤ µg/L	++	yes
CV-AAS	mono	+	ng/L	++	yes
CV-AFS	mono	+	pg/L	++	yes
ET-AAS	mono/oligo	+	< µg/L	++	yes
ICP-OES	multi	+++	< mg/L	++	yes
ICP-MS	multi	+++	< µg/L	++/+++	(yes)
DPASV	oligo	++	<µg/L	+++	(yes)
PSA	oligo	++	< µg/L	++	yes
XRF	multi	+++	mg/L (kg)	no	yes
TXRF	multi	+++	< µg/L	++	(yes)
INAA	multi	+++	< µg/kg	no	yes
RCNAA	multi	+++	≤ µg/kg	++	yes

Remarks: Potential: Mono-, oligo- or multi-element analysis possible, *Dynamic range and completeness of decomposition:* + low, ++ medium, +++ high; *Direct determination in solids and solutions:* (yes) means that this is only possible under favourable conditions, i.e. in a non-interfering analyte solution.

Abbreviations: Photo: photometry/colorimetry; F-AAS: flame AAS; H-AAS: hydride AAS; CV-AAS: cold vapour AAS; CV-AFS: cold vapour atomic fluorescence spectrometry; ET-AAS: electrothermal (mainly graphite furnace) AAS; ICP-OES: inductively coupled plasma optical emission spectrometry; ICP-MS: inductively coupled mass spectrometry; DPASV: differential pulse anodic stripping voltammetry; PSA: potentiometric stripping analysis; XRF: x-ray fluorescence spectrometry; TXRF: totally reflecting x-ray fluorescence spectrometry: INAA: instrumental neutron activation analysis; RCNAA: radiochemical (using separation procedures after irradiation) neutron activation analysis.

other liquid samples. It is also possible to analyze solids or slurries of solids directly without any previous decomposition by thermal treatment in, e.g. graphite furnaces, flames or plasmas. These approaches are being increasingly applied, particularly for materials with extremely low trace-metal contents or difficult materials and will be discussed in Sect. 11.3 of this chapter.

11.2
Error Sources

As far as contamination is concerned, the material used for the decomposition vessels has a great influence as does the proper selection of the ratio of the sample weight to the vessel surface. As already mentioned, ultrapure quartz (Suprasil) and special PTFE produced for use in medical cases (implant materials) is best suited for the purpose (see Table 10.5). If PTFE is used for the decomposition vessels it has to be carefully checked for holes and inclusions. The latter point to the fact that reworked material, often available commercially, has been used and this is definitely not suitable for the decomposition of samples with low trace-metal contents because of possible local contamination.

If materials with low levels of trace-metals have to be decomposed, all vessels used for that purpose and for storage of the resulting solutions should be scrupulously cleaned before use (see also cleaning procedures for vessels used for the collection of water samples in Chapter 4). The best procedure so far tested is the cleaning of all bottles and vessels by a steaming process in a commercially available apparatus [22, 23].

This is achieved by placing the vessels to be cleaned upside down on top of quartz tubes in an apparatus made either of glass or quartz glass (see Fig. 11.1). Vapours are passed through the quartz tubes, thus washing mainly the inner surfaces continuously. The whole procedure starts with a first washing cycle of 4–6 hours using nitric or hydrochloric acid vapours. This is followed by using water vapour for another hour.

After this procedure not only is the surface cleaned, but also the adsorption of traces of elements will be inhibited during the subsequent use of the vessel for decomposition or storage.

The purity of the reagents used for decomposition is also of particular importance. Table 10.6 demonstrates that relatively clean water and acids can be obained by distillation at just below boiling point ("subboiling distillation") in devices made of quartz or PTFE that are also commercially available.

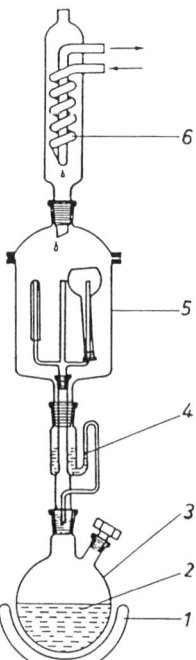

Fig. 11.1. Schematic view of a steaming apparatus. (1) heating coil, (2) nitric or hydrochloric acid, (3) round flask, (4) siphon, (5) steam chamber with the vessels that are to be cleaned, (6) reflux condenser

Sample Preparation: an Introduction

One must, however, consider that each wet decomposition results in a dilute solution. If for example an element occurs in the solid sample at the mg/kg-level, the level of the resulting analyte solution will be reduced by at least two orders (i.e. to the µg/l, level). This clearly shows the vulnerability of many solutions derived by decomposition to contamination by outside influences such as laboratory air, handling tools, as well as by the laboratory staff during the initial and all subsequent steps. This is especially the case if the elements to be determined are either very abundant in the earth's crust (e.g. Si, Al, Fe, Ca, Na, Mg) or are introduced into the environment by man-made pollution (e.g. Cu, Ni, Co, Cd, Pb). Contamination sources and techniques on how to avoid or to minimize them were already well known in the early days of trace and ultratrace analysis [26].

Thus, it is mandatory that all manipulations and the number of working steps prior to determination should be reduced to a minimum. It is also advantageous to use (quartz) decomposition vessels that allow the volume to be adjusted and the resulting solution to be stored (see Chapter 10).

If for example ultratrace analysis is performed for trace-metal levels in serum or trace elements in semiconductor materials, for reliable results all working steps from sample and reagent preparation (sometimes even from sampling, see Chapter 1) up to decomposition/determination must be performed with great care under dust-free conditions, i.e. by using clean working benches with laminar air flow or even clean laboratories [27]. Figure 11.2 shows schematically the design of such a clean laboratory.

This, relatively short overview of error sources during sample decomposition and preparation shows that decomposition procedures can introduce significant errors, which in the worst cases may reach several orders of magnitude. For some materials and elements (see Sect. 11.3) it is still nearly impossible to obtain true results if wet decomposition prior to determination is applied.

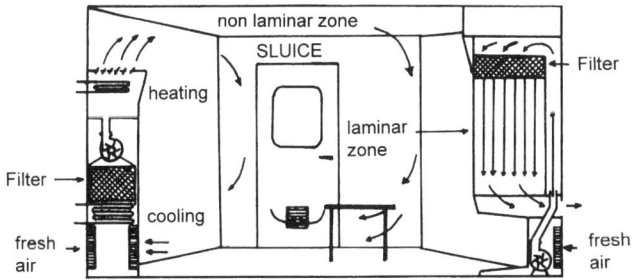

Fig. 11.2. Scheme of a clean laboratory with laminar-flow working area

11.3
Decomposition Directly Prior to Determination in Analytical Systems

11.3.1
Liquid Samples

Difficult matrices with very low endogenous trace-metal levels such as body fluids, human and animal milk, natural oils and oil products pose serious problems for reliable analysis. For these substances, modern graphite furnace techniques based on optimal background correction (Zeeman and Smith-Hieftje) systems [28] allow even the lowest natural concentrations to be analyzed routinely and without cumbersome and contamination-affected decomposition procedures. This can be achieved by maximum power heating, the choice of e.g. palladium salts [29], alternative gases such as hydrogen, freon and methane for matrix modification [30–32] and of oxygen or air for in-situ (i.e. inside the graphite tube) decomposition ("ashing") [33, 34]. Examples are the direct and accurate determination of cadmium in whole blood and urine [35, 36], of cadmium and lead in crude oil and oil products [37] as well as in human and animal milk [38].

11.3.2
Solid Samples

Since quick and contamination-free direct analysis of solid samples is highly desirable, in the past there have been numerous attempts to reach this goal by using graphite-furnace atomic spectroscopy and other spectroscopic methods. However, due to general technical problems and often insufficient background correction, these attempts were in many cases only partly successful [39, 40].

The availability of faster and more effective background correction systems, especially the commercial introduction of Zeeman background correction, has initiated an increasing number of quite successful applications of solid sampling ET-AAS as well as the analysis of aqueous slurries of solids with sufficiently small particle size by various instrumental approaches that also included flame and plasma atomization [41–46]. The present state of the art for methods based on electrothermal atomisation (ET-AAS, ETV-ICP-OES and ETV-ICP-MS) was recently reviewed by Kurfürst [47]. Both the direct analysis of solid samples and slurry sampling have the general advantage over determination after decomposition in that small sample amounts can be used (important if for example only minimal sample amounts are available or local analysis in biological structures is required) and that contamination is significantly reduced or even eliminated if no or only less extensive milling procedures (which themselves might introduce contamination) are applied. Moreover, significantly lower detection limits could be frequently obtained compared with that of solutions from wet decomposition, which is especially the case for ubiquitously occurring natural elements (e.g. the determination of Ca, K, Mg and Na by slurry ET-AAS in high-purity molybdenum trioxide [46]) and polluting elements.

Sample Preparation: an Introduction

Direct solid sampling can be manually performed by using several specially designed graphite tubes or cups as additional accessories to conventional ET-AAS systems, allowing subsamples of up to 5 mg to be taken [43, 45, 47]. There is also an AAS instrument commercially available that is specially designed for solid sampling. It uses the so-called direct Zeeman-effect (i.e. magnetic field at the radiation source) background correction [48, 49]. The instrument has proven to be very valuable for a number of applications in many different materials [45, 47]. The solid sampling applications covering all used instruments and accessories reported hitherto include the determination of up to 30 elements (Ag, Al, As, Au, Ba, Be, Bi, Ca, Cd, Co, Cr, Cu, Fe, Ga, Hg, In, K, La, Li, Mg, Mn, Mo, Ni, Pb, Rb, Se, Si, Tl, Tl and Zn) in a large number of biological, environmental, geological, polymer and industrial materials [43, 45, 47].

The limitations of this technique are certain difficulties with calibration (either by using certified reference materials or reference solutions) and that automated sample introduction is still not possible. There is at present only a semi-automat-

Table 11.2. Comparison of properties of slurries and solid samples [50]

Slurries	Solid samples
Subsample mass: typical 0.02 mg (effective introduced mass)	*Subsample mass:* Sample mass usually typically 1 mg (effective introduced mass)
Representative mass: up to 50 mg with one slurry preparation (element extraction)	*Representative mass:* typically 10–20 mg (number of replicates times mass of sub-samples)
Normal AAS instruments applicable	Special AAS-systems (tubes, boats) required
Automation commercially available and even in laboratory made systems relatively simple to achieve	Complete automation difficult; at present only a semiautomated device commercially available
Dilution easy	Dilution difficult for high metal levels
Matrix modification easy	Matrix modification not so easy
Determination of refractories easy	Determination of refractories not easy
Ground samples only	All types of samples applicable
Elevated contamination risk due to the use of acids, thickeners etc. and matrix modifiers	No or very low contamination risk if no reagents are used
Minimum representative mass at the mg level because of the small injected volume of approx. 20 µl and subsample mass down to 0.02 mg	Extremely small sample masses and single samples (organisms, parts of plants etc.) with weights at the µg level applicable
Analyte extraction from the preparations makes a reliable homogeneity determination only possible, if the percentage of extraction is 100% or exactly known. The latter is often difficult to obtain	Homogeneity determinations easy with direct solid sampling, because of the simple use of the effective introduced mass for calculations of the homogeneity factor
Detection limits at the order of that for liquids, in some cases somewhat lower	Very low relative detection limits attainable compared to liquids and sometimes also to slurries

ed accessory available for the commercial solid sampling system. This is because there are significant problems for fully automating the introduction of solid materials into atomization systems. See also Table 11.2 showing the virtues and limitations of direct solid and slurry sampling.

Slurry sampling requires particle sizes small enough to pass through the transport tubing and injection parts without hindrance and that the specific weights or ranges of specific weights of the materials to be analyzed are small enough to allow the formation of suspensions that will remain stable during the necessary procedural steps. If predominantly ultrasonification for complete mixing before sample injection is used, together with a commercially available system [44, 50, 51] and quite similar laboratory-made systems, automated operation of slurry sampling is possible with nearly all commercially available ET-AAS systems [52]. This, together with less difficult calibration compared to direct solid sampling, might be the reason that more papers have now appeared dealing with slurry sampling ET-AAS. Very often the particle sizes of the materials to be analyzed are sufficiently small for introduction into ET-AAS systems if the tube diameters are slightly increased. The determination of at least 25 elements (Al, As, B, Be, Ca, Cd, Co, Cr, Cu, Fe, K,

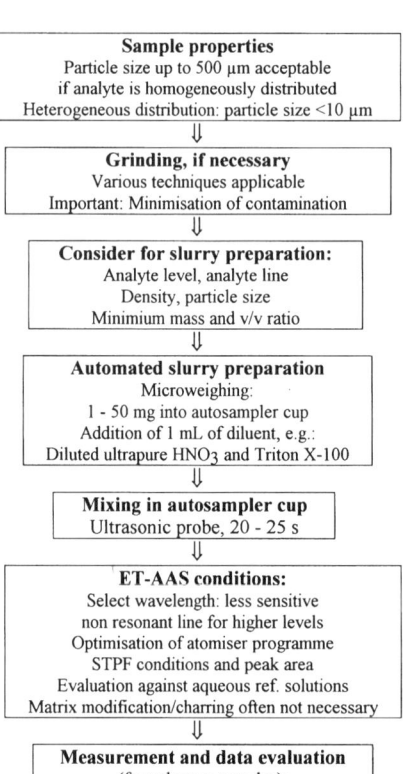

Fig. 11.3. Flow chart for the whole slurry sampling procedure [after 52]

Li, Mg, Mn, Mo, Na, Ni, Pb, Pd, Se, Si, Ti, Tl, V and Zn) in biological, environmental, gelogical, polymer and industrial materials has been reported [52].

Disadvantages of this approach are the need for the addition of reagents (e.g. acids and some thickeners to maintain slurries in suspension) and the use of sophisticated sample stirring devices, transport and injection accessories, so that the introduction of contamination might be easier than with direct solid sampling. For the introduction of slurries into flames and plasmas, however, much smaller particle sizes than for ET-AAS are necessary. This needs special, often vigorous, grinding procedures down to a few μm that may introduce contamination for difficult elements. Further differences between the two approaches are shown in Table 11.2. Figure 11.3 gives a general scheme for the preparation and analysis of slurries that is partly also applicable to direct solid sampling.

The often-expressed fear that both approaches, because of the low sample amount used, might not be accurate enough has been disproved by many comparisons with conventional procedures [45], by analyzing certified reference materials as well as by very acceptable results from international collaborative studies [53, 54].

11.4
Quality Assurance

The reliability of each decomposition procedure should be checked with materials of proven homogeneity and element content (see also Chapter 10). This can be best accomplished in general by the use of appropriate certified reference materials, in-house reference and control materials and carefully and continuously performed statistical control procedures [55–58]. Appropriate means that the material used to check a distinct decomposition procedure or series of determinations should be as close as possible in contents of matrix, main and trace elements [59]. The variety of certified reference materials produced either by official institutions or private companies currently on offer can be obtained from a number of commercial distributors based on almost yearly editions of available materials, listed for example in the COMAR database [60], literature surveys such as [61] and catalogues of CRM producers such as that from the European Commission [62].

11.5
References

1. Bernhard M, Brinckman FE, Sadler PJ (eds) (1986) The importance of chemical speciation in environmental processes. Springer, Berlin Heidelberg New York
2. Ure AM (1990) Fresenius J Anal Chem 337:577–581
3. Broekaert JAC, Gücer S, Adams F (eds) (1990) Metal speciation in the environment. Springer, Berlin Heidelberg New York
4. Batley GE (ed) (1990) Trace element speciation: analytical methods and problems. CRC Press, Boca Raton, FL
5. Donard O, Quevauviller P (eds) (1992) Proc Arachon workshop on speciation. Mikrochim Acta 109:1–190
6. Chakrabarti CL, Lu Y, Cheng J, Back MH and Schroeder WH (1993) Anal Chim Acta 267: 47–64

7. Donard OFX, Ritsema R (1993) Hyphenated techniques applied to the speciation of organometallic compounds in the environment. In: Barceló D (ed) Environmental Analysis. Elsevier, Amsterdam, p 550
8. Van den Berg CMG (ed) (1994) Special issue on metal speciation. Anal Chim Acta 248:461–668
9. Quevauviller P (1994) Appl Organomet Chem 8:715–725
10. Dunemann K, Begerow J (1995) Kopplungstechniken zur Elementspurenanalytik. VCH Publisher, Weinheim
11. Lin Y, Smart NG, Wai CM (1995) Supercritical fluid extraction and chromatography of metal chelates and organometallic compounds. trac 14:123–132
12. Gardiner PHE (1987) Topics in Current Chemistry 141:145–174
13. Cornelis R, Borguet F, De Kompe J (1993) Anal Chim Acta 283:183–189
14. Chen D, Luque de Castro MD, Valcáral M (1991) Determination of anions by flow injection – a review. Analyst 116:1095–1111
15. Crompton TR (1996) Determination of anions. Springer Berlin Heidelberg London New York
16. Mamba S, Kratochvil B (1995) Intern J Environ Anal Chem, 60:295–302
17. Stoeppler M (1992) Analytical methods and instrumentation – a summarizing overview. In: Stoeppler M (ed) Hazardous metals in the environment. Elsevier Science Publishers B.V., Amsterdam, p 97
18. Dogan S, Membrini G, Haerdi W (1981) Anal Chim Acta 130:385–390
19. Warren HV, Hoksy SJ, Gould CE (1983) Sci Total Environ 29:163–169
20. Stoeppler M, Apel M (1984) Fresenius Z Anal Chem 317:111–116
21. Bock R (1979) A handbook of decomposition methods in analytical chemistry. Int Textbook Co Ltd, London
22. Tschöpel P (1980) Aufschlußmethoden. In: Ullmann's Encyclopädie der Technischen Chemie. Verlag Chemie, Weinheim, Vol 5, p 27
23. Tschöpel P (1992) Sample treatment. In: Stoeppler M (ed) Hazardous metals in the environment. Elsevier Science Publishers B.V., Amsterdam, p 73
24. Sansoni B, Panday VK (1994) Sample treatment of human biological materials. In: Herber RFM and Stoeppler M (eds) Trace element analysis in biological specimens. Elsevier Science Publishers B.V., Amsterdam, p 21
25. Subramanian KS (1996) Spectrochim Acta 51B:291–319
26. Zief M, Speights R (1972) Ultrapurity, methods and techniques. Marcel Dekker Inc, New York
27. Boutron CF (1990) Fresenius J Anal Chem 337:481–491
28. Robinson JW (1990) Atomic Spectroscopy. Marcel Dekker Inc, New York
29. Schlemmer G, Welz B (1986) Spectrochim Acta 41B:1157–1185
30. Frech W, Cedergren A (1976) Anal Chim Acta 82:93–102
31. Welz B, Schlemmer G (1988) At Spectrum 9:76–80
32. Welz B, Schlemmer G (1988) At Spectrum 9:81–83
33. Beaty M, Barnett W, Grobenski Z (1980) At Spectrum 1:72–77
34. Eaton DK, Holcombe JA (1983) Anal Chem 55:946–950
35. Herber RFM, Stoeppler M, Tonks DB (1990) Fresenius J Anal Chem 338:269–278
36. Herber RFM, Stoeppler M, Tonks DB (1990) Fresenius J Anal Chem 338:279–286
37. Narres HD, Mohl C, Stoeppler M (1984) Int J Environ Anal Chem 18:267–279
38. Narres HD, Mohl C, Stoeppler M (1985) Z Lebensm Unters Forsch 181:111–116
39. L'vov BV (1976) Talanta 23:109–118
40. Langmyrh TJ (1979) Analyst 104:993–1016
41. Langmyrh TJ, Wibetoe G (1985) Prog Analyt Atom Spectosc 8:193–256
42. Langmyrh TJ (1985) Fresenius Z Anal Chem 322:654–656
43. Bendicho C, de Loos-Vollebregt TC (1991) J Anal At Spectrom 6:353–374
44. Miller-Ihli NJ (1992) Anal Chem 64:964A–968A
45. Kurfürst U (1990) Die direkte Analyse von Feststoffen mit der Graphitrohr-AAS. In: Günzler H et al. (eds) Analytiker Taschenbuch, Band 10, Springer, Berlin Heidelberg New York, p 190

46. Docekal B, Krivan V (1993) J Anal At Spectrom 8:637–641
47. Kurfürst U (ed) (1996) Solid Sample Analysis – Direct and Slurry Sampling using ET-AAS and ETV-ICP. Springer (in press)
48. Kurfürst U (1983) Fresenius Z Anal Chem 315:304–320
49. Rosopulo A, Grobecker KH, Kurfürst U (1984) Fresenius Z Anal Chem 319:540–546
50. Epstein MS, Carnrick GR, Slavin W, Miller-Ihli NJ (1989) Anal Chem 61:1414–1419
51. Miller-Ihli NJ (1992) Atom Spectrosc 13:1–6
52. Stoeppler M, Kurfürst U (1997) Slurry sampling into the electrothermal atomizer. In: Kurfürst U (ed) Solid Sample Analysis – Direct and Slurry Sampling using ET-AAS and ETV-ICP. Springer, in press
53. Herber RFM, Grobecker KH (1995) Fresenius J Anal Chem 351:577–582
54. Miller-Ihli MJ (1995) Spectrochim Acta 50B:477–488
55. Griepink B, Stoeppler M (1992) Quality assurance and validation of results. In: Stoeppler M (ed) Hazardous metals in the environment. Elsevier Science Publishers B.V., Amsterdam, p 517
56. Braithwaite RA (1994) Interlaboratory and intralaboratory surveys. Reference methods and reference materials. In: Herber RFB and Stoeppler M (eds) Trace element analysis in biological specimens. Elsevier Science Publishers B.V., Amsterdam, p 213
57. Parr RM, Stoeppler M (1994) Reference materials for trace element analysis. In: Herber RFM and Stoeppler M (eds) Trace element analysis in biological specimens. Elsevier Science Publishers B.V. Amsterdam, p 233
58. Herber RFM, Sallé HA (1994) Statistics and data evaluation. In: Herber RFM and Stoeppler M (eds) Trace element analysis in biological specimens. Elsevier Science Publishers B.V., Amsterdam, p 257
59. Jenks PJ (1995) Fresenius J Anal Chem 352:3–4
60. Klich H, Pradel R (1995) Fresenius J Anal Chem 352:23–27
61. IAEA Vienna (1995) Survey of reference materials. Vol 1: Biological and environmental reference materials for trace elements, nuclides and microcontaminants. IAEA-TEC-DOC-854
62. IRMM (1996) BCR Reference materials, Institute for Reference Materials and Measurements (IRMM), Retieseweg, B-2440 Geel, Belgium

Chapter 12
Pressure Digestion: Apparatus, Problems and Applications

Ewald Jackwerth and Michael Würfels
Translated by Philip H. E. Gardiner

12.1 Introduction

The expression "pressure digestion" is a misnomer because the impression given is that pressure is critical for the digestion process. Indeed, it is during the digestion reaction that pressure builds up, and the role this plays in the digestion process will be explained later. However, it is important to note that the pressure build-up, amongst other things, does introduce some dangers in the application of these methods. These methods are better described if they are grouped under the title, "Wet digestion methods with the application of solubilising and/or oxidising digestion reagents in closed systems".

The forerunner of these methods was first described in 1860 by G. L. Carius, who described the digestion of organic materials with concentrated nitric acid at 250–300 °C [1, 2]. The sample and the acid were mixed in a thick-walled quartz ampoule and sealed. The ampoule was transferred into a "bomb canister" and heated in what was called a "bomb oven" for several hours, after which it was cooled, opened, and the contents analysed. The choice of military words to describe the apparatus points to the dangers of the careless use of the Carius-digester.

The first demountable external jacket system was made from stainless steel; however, those used for mineral analysis had an inner coating of platinum. Liners made from poly(tetrafluoroethylene) (PTFE) have been used since 1955 [3–12]. The contributions of B. Bernas (1968) and G. Tölg and co-workers (1972) are worth mentioning with regard to the commercialisation of the decomposition vessels or "digestion bombs" as they are often called. Today, there are a number of systems covering the whole market range, however, these will not be described individually or any comparisons made. In this chapter, typical equipment components, digestion parameters, safety and the analytical capabilities of the systems will be discussed.

Although there are differences in construction details, there are similarities in a number of components in the various digestion systems, and a schematic diagram of a typical system is shown in Fig. 12.1. At the heart of each system is the insert or liner, made mainly from inert material, in which the digestion reaction occurs. The liner is encased by a container that can withstand the type of pressures developed during digestion. The container is closed via a compression spring, which presses against the lid so that the liner is tightly closed. The compression spring acts as a pressure release valve, particularly in reactions which generate lots of gases and hence generate relatively high pressures. Most systems incorporate a breakable screw in the base plate which serves to release pressure when broken and

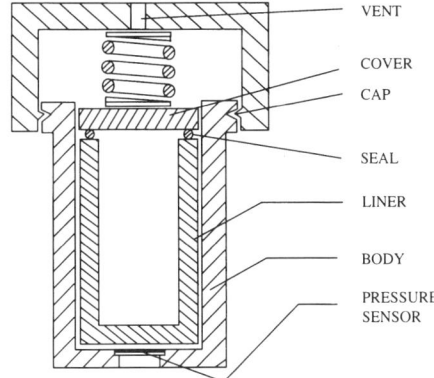

Fig. 12.1. Pressure decomposition system

thus prevents explosions. Systems are temperature controlled by using an appropriate heating block or oven which is heated to a preselected digestion temperature and maintained at that temperature via a thermostat.

In current practice the following criteria are typical for digestion in closed systems.

- The dimensions of the liner and sample mass must be compatible. Especially with organic substances, a maximum sample mass of about 100 mg should be used.

Highest purity commercially available or isothermally distilled nitric acid should be used for organic materials [13]. Other mineral acids, often in combination with hydrofluoric acid, are suitable for inorganic substances.

Note that there is the likelihood of an explosion when perchloric acid is used on its own or even in a mixture of acids [14–16].

- Temperatures that are much higher than the boiling point of nitric acid are attainable with the closed system. As a result, the oxidising potential of the acid, and the reaction rate are greatly enhanced. Consequently the digestion time in comparison with an open system is much reduced. Furthermore, the range of materials that can be digested with nitric acid is extended.

Excess digestion reagents and intermediate products from the acid and sample constituents (NO_2 and free radicals) are contained in the decomposition vessel, and as a result the oxidising rate is faster, and the digestion goes further to completion. Volatile trace elements, and their various species, are retained as long as diffusion (possible through the walls of the container) is prevented [17].

- Blank levels are low because of the reduced amount of reagents, and the effective exclusion of external contamination. The high mean blank value $\bar{x}_{Bl}$ and the standard deviation of the results s_{Bl} are directly influenced by the measured detection limit $\underline{x}$ and are a function of the calculated detection limit $\underline{c}$ [18]. The equations linking the various quantities are given below:

$\underline{x} = \bar{x}_{Bl} + 3s_{Bl}$

$\underline{c} = f(x)$

In order to ensure long-term reliability and safe operation it is essential that careful consideration is given to the properties of the various materials used for the construction of the parts of the digestion vessel, and their compatibility with the chemical and physical behaviour of the reactants [19]. Germany, for example, has regulations governing the safe use of pressure decomposition systems [20, 21].

12.2
Liner

The essential properties of the liner vessel are:

- good resistance against attack from the digestion mixture, and any products that may result from the digestion of the different materials,
- high purity so that tolerable blank levels are obtained when the surface of the container is attacked,
- low gas permeability and a negligible sorption potential for trace substances, and
- thermal and dimensional stability at the digestion temperature.

Of the range of materials that have been tested using the above criteria, only a few have come close to fulfilling them. These belong to the group of high density polymers of which polytetrafluorethylene (PTFE) is prominent, and more recent additions include fluorinated ethylene propylene (FEP), and perfluoroalkoxy fluorocarbons (PFA). The chemical resistance of these materials is almost identical; however, they differ in hardness, translucence, mechanical and thermal properties, and workability.

In a number of systems, alternative containers, made either from quartz glass or glassy carbon, have been used [25]. These offer few advantages, and are only useful when the containment acheved by PTFE is inadequate. PTFE is the material of choice because of its chemical resistance, and its ability to withstand high temperatures of up to 260 °C. The common digestion reagents (nitric and hydrofluoric acids) attack PTFE when they are used regularly. However, only small amounts of the surface are removed. PTFE is, to a large extent, resistant to attack from other concentrated acids. Fluorine, extremely reactive fluoride-containing species and molten alkali metals attack PTFE at high temperatures [22, 24].

In addition to chemical resistance, it is essential, particularly for trace analysis, that the PTFE material is of the highest purity. Impurities result not only from the manufacturing process but also whilst working the materials into shape. It is therefore necessary to treat these containers with the digestion reagents [13, 26] until acceptable blank levels are obtained. It is noteworthy that contaminants in the polymers that are not removed by acid treatment may, during the lifetime of the vessel, be set free, and thus lead to inexplicably high blank values. For this reason, it is essential that the PTFE material used for this purpose is of the highest purity, which is equivalent to that used for the manufacture of prosthesis implants.

One of the typical properties of PTFE and other polymers is the relatively high gas permeability [23]. Consequently, during digestion acid vapour and nitrogen dioxide, as well as gaseous reaction products and traces of the sample material could diffuse through the walls of the liner. At trace levels these losses could lead to low recoveries. It is also important to note that the carry-over from substances that have diffused into the liner walls could lead to the contamination of subsequent samples. This is known to occur with Hg and I_2 [12, 26, 27]. However, systematic studies of the adsorption of traces of substances on PTFE, have shown that losses are negligible [11, 28] for determination at $\mu g\ g^{-1}$ and higher $ng\ g^{-1}$ levels.

The mechanical behaviour of PTFE is of importance in its use in pressure digestion systems. This characteristic is influenced by the extent of cross-linking and crystallinity of the material. A reversible volume increase of up to 25% [22, 24, 29] can occur when PTFE is heated to its sinter temperature (370–380 °C). The deformability of PTFE under pressure and high temperature results in the closing up of the gap between the liner and body. Consequently, the design of the pressure release system must make allowance for the fact that the body could be completely filled when the maximum allowable temperature is attained. In addition, it is also essential that the function of the pressure release valve is not compromised under these conditions. Furthermore, the liner should still be tightly sealed when the system is cooled to room temperature. PTFE made by isostatic pressing has thicker walls and as a result expands less at high temperatures. This form of PTFE is less porous and expands by between 3–4% at typical digestion temperatures.

The mechanical stability of the polymer is adversely affected when temperatures greater than 200 °C are used, although PTFE thermally decomposes at about 400 °C. High temperatures make PTFE more permeable to gases and also the useful life of the liner is reduced. In the majority of applications, particularly those which involve organic materials, temperatures between 170–180 °C have been found to be adequate. At higher temperatures, the excessive pressures generated are transferred to the lid. However, in cases where springs are used in the pressure release valve mechanism, damage to the lid and container from the force exerted on the spring is minimal. In contrast, permanent damage to the lid and insert could result when other pressure release valve systems are used. PTFE-O-ring seals are equally effective in providing a seal, and in addition they are cheap to replace.

Another important design consideration is how to minimise the risk of expelled hot digest solutions coming into contact with the metal casing.

For a given liner volume the pressure release valve system should be such that it can operate safely at the pressures generated in normal use. Systematic investigations have shown that the liner volume can be regarded as optimum when, during the digestion of organic materials, a pressure of about 20 bar is generated, and the carbon content of the digest is between 1.5–3 mg C/ml. For the digestion of organic materials, liner volumes of between 25–40 ml, and sample masses between 100–300 mg have been found to give optimum results [17, 30]. Larger volumes reduce the digestion rate, and in smaller volumes excessive pressures are generated, leading to the loss of sample digests through venting.

12.3
Body

Bodies encasing the liners are made from high grade stainless steel in most commercial systems. Consequently, high Fe, Cr and Ni blanks could result when the digest solution comes into contact with parts of the body. This contamination risk can be reduced by coating the surface of the body with PTFE. However, corrosion of the casing could still occur, particularly when nitrous gases or hydrochloric acid vapour are given off either during the release of excess pressure or when the vessel is opened. Corrosion products can be removed by rubbing the body with concentrated phosphoric acid [25].

Differences in the construction of the body and cover determine the behaviour of the gases that are produced during digestion, particularly when there is still residual pressure on cooling the system. Systems in which the cap is screwed onto the body are particularly difficult to open when there is residual pressure. This situation is worsened when the body is corroded. In such a case, unscrewing the cap may require the use of a tool. In other systems the residual pressure can be released by pressing the vent spring on the cap. This enables the cap to be easily unscrewed from the body.

There are commercial systems which are designed to take more than one digestion vessel in a single body and descriptions can be found in the literature [31, 32].

12.4
Safety Devices

The course and extent of the pressure increase during digestion is determined by the sample type and mass. Pressures of up to 10 bar [19, 33] can be generated during digestion. The safety valve with a spring mechanism is designed to apply the necessary pressure so that a tight seal is achieved, and also to relieve any excess pressure. It is therefore important to ensure that the performance of the spring is not compromised with use. Consequently, the vent and spring should be examined regularly and if necessary replaced. The design of the vent system should be such that its function is not impaired by damaged PTFE parts, and in addition the expulsion of excess hot acid should be directed away from the metal body.

Delayed oxidation reactions have been observed when organic materials are digested with nitric acid. Inhibition of the reactions can occur despite the increase in temperature, followed by a sudden build-up of pressure in the vessel as the reaction rate increases [33]. A similar effect has been observed with substances which form two phases with the digestion acid such as powders and oil samples. The level of the pressure peak and the elasticity of the spring system determine whether the sudden generation of high pressures within the system lead to the expulsion of the digestion solution through the safety vent. It is important that the pressure membrane is designed to withstand the sudden pressure surges and also to prevent damage to the liner and expulsion of the contents. In the event of an explosion, it is likely that parts of the decomposition and heating system, and contents would be propelled by a powerful force. It is therefore essential that unauthorised personnel are prohibited from entering the laboratory area in which pressure decomposition is carried out [21].

Consideration should be given to the fact that explosive mixtures may be formed when organic substances are digested with concentrated nitric acid. The risk is increased in cases where nitric acid and concentrated sulphuric acid are used together. Attempts to digest fat and oil have led to massive explosions with the result that the digestion vessels have been destroyed [33–35]. The pressure wave generated by the detonation is so fast that the spring valve does not open in time to relieve the excessive pressure. Studies have been carried out on the stability of commercially available systems with known quantities of explosives [36].

In order to reduce the extent of any explosion resulting from unpredictable reactions, it is essential that the sample mass used should be between 100–300 mg. Reduced sample mass and the possibility of remotely switching off the heating block in an emergency are useful precautions.

12.5
Heating System

The purpose of the heating system is to bring the enclosed digestion vessel rapidly (but with a uniform temperature distribution throughout the digest) to the digestion temperature, usually between 170–180 °C, and to thermostatically keep the temperature constant for a few hours. At the end of the heating cycle, the system may be cooled by air or water circulation. Besides time- and temperature-regulated ovens, there are also heating blocks, which can accommodate more than one digestion vessel and are heated via hot plates or heatable magnetic stirrers. The temperature of these devices is controlled by temperature sensors. The use of a heating block provides additional casing, and from a safety point of view this is to be preferred.

12.6
Conditions for Digestion

The important digestion parameters such as sample mass, quantity and type of digestion mixtures, temperature and time, can only be varied within a narrow range of values as has been shown by a number of systematic studies. This applies particularly to the decomposition of organic samples. As already indicated, for safety reasons, the sample mass should be between 100–300 mg. The use of 65% w/v HNO_3 is recommended, and in most applications between 1–2 ml is adequate. Furthermore, the maximum allowable temperature when using PTFE liners is 180 °C. For the digestion of biological materials temperatures between 170–180 °C have been found to give optimum decomposition [37], suitable for most purposes. Note that experiments in which temperatures of up to 220 °C were used did not produce any significant improvement in the carbon content of the digest. A selection of pertinent publications can be found in the literature [11, 25, 30, 31, 38–49].

In order to optimise the conditions for the pressure digestion of biological samples with nitric acid, a number of investigations have been carried out [17, 30, 37, 50, 51]. The findings from these can be summarised as follows:

Digestion time: Three hours has been found to be optimum when the digestion temperature is between 170–180 °C, irrespective of the biological sample type [37]. No change in content of the residue is obtained if longer digestion times are used.

Acid volume: Two ml of HNO_3 is adequate for the digestion of biological materials provided the carbon content of the samples is about 100 mg, and digestion temperatures of between 170–180 °C are applied for 3 hours [17, 37].

Table 12.1. The carbon content of freeze-dried biological materials

Sample type	Carbon content (%)	Sample type	Carbon content (%)
Plant material		Liver	51
Brown algae	35	Fish fillet	52
Wheat	45	Whole blood, human	52
Spinach	38	Whole blood, pig	52
Poplar leaves	44	Oyster flesh	46
Beech leaves	48	Pig's kidney	49
Maple leaves	48	Mussel tissue	41
Grass	39	Hen's egg	50
Clover	36	*Pure carbohydrate*	
Pine needles	51	Glucose monohydrate	37
Spruce needles	48	Cane sugar (saccharose)	42
Spruce shoot	49	Lactose	42
Spruce bark	36	Cellulose	43
Peach (fruit flesh)	40	Starch	48
Animal material		*Fat*	
Skimmed milk powder	42	Butter, vegetable oil, vegetable fat, tallow	74–78
Full cream milk powder	52		
Lean beef	50		

Sample mass: Consider a digestion system with a PTFE liner volume of 35 ml which can withstand a pressure of 20 bar at the optimum digestion of temperature 180 °C. A sample with a pure carbon content of 100 mg can be digested under the above conditions. A pressure build-up of 20 bar due to the production of CO_2, NO/NO_2 and water can occur [26]. It is possible to calculate the mass of biological sample from the carbon content. The carbon content in freeze-dried materials is summarised in Table 12.1. It is important to take into consideration the water content, and to use a higher mass when dealing with fresh samples. Examination of Table 12.1 reveals that the following masses are adequate: 200 mg for freeze-dried tissue (ca. 50% C), 200–300 mg for freeze-dried plant material, (30–50% C) and only 120 mg for pure fat samples (78% C).

Note that for optimum digestion the ratio of sample mass (relative to the carbon content) to the liner volume should not exceed 1.5 mg C/ml so that there is an excess of active NO/NO_2 [17]. Without the slight excess of NO_2, the extent of the digestion, measured by the conversion of carbon to carbon dioxide, would be adversely affected.

It is difficult to give a generalised recommendation for the digestion of inorganic materials. It is essential to digest the samples with an oxidising agent or a mixture of two reagents. Nitric acid or a mixture with sulphuric acid and hydrochloric acid can be used as it is very unlikely that decomposition of these materials could cause explosions. Other reagents that have been used include phosphoric acid, hydrogen peroxide, and bromine. The use of perchloric acid requires extreme caution as explosive mixtures can form. A solution of 40% hydrogen fluoride is recommended for the initial digestion of silicates followed by a mixture of acids. The addition of boric acid to the digest prevents the precipitation of poorly soluble metal fluorides [9, 52]. More than 1 g of samples, which present no risk of explosion, can be digested.

Digestion times between 0.5 to 3.5 hours have been quoted in the literature [11, 25] as optimal for both organic and inorganic samples. Not only does the digestion time depend on the type of sample material but also on the heat capacity of the metal encasement of the liner. Therefore in investigations the heat capacity of the encasement should always be quoted. The rate of digestion of inorganic material can be greatly increased by introducing the samples in the form of powders and by stirring the reaction mixture. A selection of the procedures for the digestion of minerals with different mixtures have been published [4, 10, 32, 38, 52–62].

12.7
Completeness of Sample Decomposition

Since the introduction of pressure decomposition to trace element analysis, it has been shown in publications that digestion of some organic substances with HNO_3 at 180 °C does not produce a completely mineralised digest [17, 19, 25, 30, 33, 37, 50, 63–65]. In these cases extending the digestion time and increasing the quantity of nitric acid does not improve the extent of digestion. The presence of residual matrix constituents could lead to interference effects particularly when the digest is analysed by voltammetry. Such interference effects could lead to signal depression or the appearance of peaks which interfere with the analyte signal [25, 51, 63, 66–71].

A number of investigations [17, 30, 37, 50, 51, 64, 70] have examined the factors that affect the completeness of the digestion of biological samples with nitric acid, and have identified reactants that are qualitatively converted by nitric acid and the products of the reactions.

Using the following conditions: 180 °C, 3 hours, 2 ml 65 % HNO_3 per 100 mgC, the ratio of sample mass to liner volume ≥ 1.5 mgC/ml clear digest solutions are obtained for all types of biological samples. However, some samples contain silicates, for example plants and mussels, and these can be dissolved with a mixture of nitric acid and hydrofluoric acid [17]. Undissolved species such as calcium fluoride and fluorosilicates can be dissolved by bringing the contents of the digestion vessel to dryness, and adding nitric acid to convert the fluorides to soluble nitrates [17]. Depending on the sample type, the presence of silicate could be tolerated, provided it can be established that the trace elements of interest are not bound to the insoluble precipitate. The need to eliminate the use of hydrofluoric acid is particularly important when sample digestion is carried out in quartz liners.

Experience has shown that organic matrix constituents, and their reaction products with nitric acid, can be brought completely into solution. In a majority of biological materials, the constituents are converted fully into CO_2, H_2O and volatile products, which are lost at 120 °C, and the carbon content of the residue is undetectable coulometrically. However, organic residues, which are not attacked by nitric acid at 180 °C, have been found [17, 51] when some protein-containing materials and some fats are digested. Pure carbohydrates (sugars, starch, cellulose, etc.) are mineralised with nitric acid at 180 °C [17, 37]. The trace metal content of the resultant digest can be determined without difficulty by either inverse voltammetry or any other suitable analytical technique. Carbon-free solutions are also obtained when fats, which do not contain multiple unsaturated fatty acids, are digested. If the fats contain either linoleic or linolenic esters, then on addition of acid, linoleic and linolenic acids are formed respectively, and these are converted to 1,2 cyclopropandicarboxylic acid, which at 180 °C is not attacked by HNO_3 [51]. Although the presence of 1,2 cyclopropandicarboxylic acid does not interfere with elemental electrochemical determinations [17, 51], adverse effects on other methods cannot be ruled out completely.

The digestion of protein- and amino acid-containing samples with nitric acid leads to the formation of electrochemically active nitrated benzoic acid from phenylalanine. Consequently, the determination of Cd, Cu, Pb and Zn by inverse voltammetry suffers from several signal interferences, and in some cases the extent of the interference is so severe as to preclude their analysis [17, 51]. Besides nitrobenzoic acid, other amino acid breakdown products could be formed [17, 51] and as observed in the digestion of linoleic and linolenic acid-containing samples these could also cause interference effects.

Of all the techniques for the determination of trace elements, atomic spectrometry is the least affected by the presence of organic residues. However, these trace residues can affect the performance of different atomic spectrometric techniques. Their presence can change the nebulisation rate, through their influence on the formation of the aerosol in the nebuliser and the vaporisation of trace species. Such trace substances are also a source of background signal in carbon-furnace atomic spectrometry. Furthermore, when aqueous standards are used for instrument calibration, systematic errors, which are difficult to correct, could result. The use of matrix-matched certified reference materials is invaluable for establishing the extent of such systematic errors [72]. Besides the need to limit the sample mass to about 100 mg, incomplete digestion of a few sample types is a major disadvantage in the pressure decomposition of biological samples in PTFE liners.

One of the causes of incomplete decomposition of organic materials is the low oxidation potential of nitric acid at 180 °C. Increasing the digestion time and the amount of nitric acid used does not, as already indicated, lead to better results. In this respect, the behaviour of organic substances is different from inorganic materials, which are hard to digest in acid. For these, increasing the amount of acid, and the use of longer heating times with stirring, does significantly improve the extent of digestion. With the digestion of organic materials small quantities of perchloric acid could be used to destroy any organic residues. However, the presence of perchloric acid could result in higher blank values, and necessitate the use of extraction systems; moreover, extreme caution has to be exercised when using this acid because of the explosion risk.

Temperatures above 300 °C are required for the complete digestion of organic materials when nitric acid alone is used in the closed system. Such high temperatures are unattainable with the system described here. However, Knapp [73] has described a commercially available system with which it is possible to achieve complete digestion without the need to add perchloric acid. The equipment consists of a stainless steel autoclave and a sealed quartz digestion vessel. The vapour pressure generated on heating the digestion mixture is counterbalanced by applying an external pressure of 100 bar, which is generated by filling the autoclave with nitrogen.

Regardless of the type of biological material, carbon-free digests are obtained in which the various elements are present as free ionic species and which are universally compatible with the sampling requirements of various analytical techniques. The digests can be analysed by both inverse voltammetry and hydride-generation atomic absorption spectrometry, two techniques whose performance can be impaired if the digestion is incomplete. Note that species of elements such as As, Hg, Se and Sb which form stable bonds with carbon may not be completely destroyed even at 300 °C. For example, the chemically very stable compound arsenobetaine, found predominantly in marine samples, requires a temperature of 320 °C for complete decomposition (cf. Chapters 10 and 14).

Figure 12.2 shows signals for Cd, Cu, Pb and Zn obtained when algae (NIES Reference Material, *Sargassum Fulvellum*, No. 9) and mussel (NIES Reference Material, *Mytilus edulis*, No. 6) are analysed by square-wave voltammetry. The digests were evaporated to dryness, in order to expel any oxides of nitrogen, which interfere with the generation of the reduction signal [17, 70]. The background currents are similar for both the digests and the aqueous standards. The voltammograms are interference-free, and the area under each peak is easily determined. These observations also apply to determinations by inverse differential pulse voltammetry.

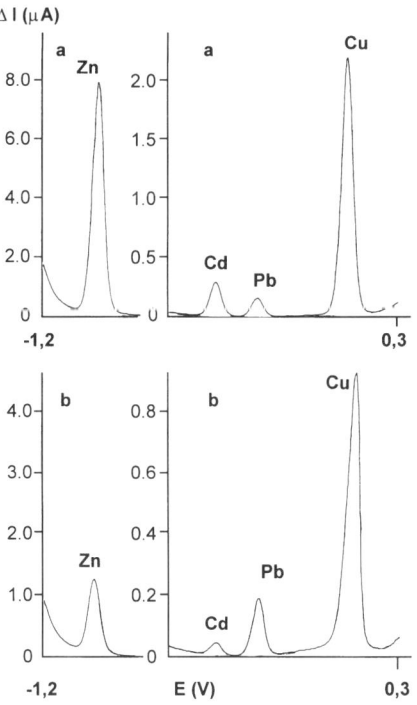

Fig. 12.2. Square-wave voltammograms of digest solutions from (a) the mussel *Mytilus edulis*, and (b) algae, *Sargassum fulvellum*, after nitric acid pressure digestion at 300 °C

The autoclave-based digestion procedure has been tested on a number of biological samples, and interference-free voltammograms have been obtained in all cases. Besides Cd, Cu, Pb and Zn, other elements such as As, Co and Ni have been determined in most biological matrices after wet digestion at $\geq 300\,°C$ without the need to treat the samples with the potentially explosive mixture of $HClO_4/H_2SO_4$. Interference-free determination of Co and Ni by adsorption voltammetry and As by hydride-generation atomic absorption spectrometry have been achieved.

The advantages of this closed digestion system include:

(i) the danger of working with perchloric acid is removed as most biological materials and foods can be digested,
(ii) the digest is suitable for analysis by any elemental determination technique, and
(iii) the reliability of the elemental analysis is improved because a major source of interference, namely the organic residue is completely oxidised to CO_2 and eliminated. As a result, the precision and accuracy of the analysis is vastly improved in comparison to existing wet digestion procedures.

It is important to note that carbon-free digestions are obtained when biological samples containing 100 mg carbon in 30 ml quartz liners with 2 ml of conc. HNO_3 in an autoclave set at 320°C, corresponding to 300°C in the digestion vessel, is digested for 1–2 hours [17].

In conclusion, closed-vessel pressure digestion as described here is the technique of choice for the vast majority of biological materials and food samples. The technique is fast, reliable, reproducible and safe when used properly. Furthermore, in cases where the digestion is complete, improved recoveries are obtained, and the digests are compatible with most elemental determination techniques. However, the mode of heating the digestion vessels is slowly being supplanted by microwave technology and this is the subject of the subsequent chapter in this book.

12.8
References

1. Carius GL (1860) Ann Chem 136:1
2. Carius GL (1870) Ber Dtsch Chem Ges 3:697
3. May J, Row JJ (1965) Anal Chim Acta 33:648
4. Lounamaa K (1955) Fresenius Z Anal Chem 146:422
5. Ito J (1962) Bull Chem Soc Japan 35:225
6. Wahler W (1964) Neues Jahrb Mineral Abhandl 101:109
7. Langmyhr FJ, Graff PR (1965) Norg Geol Undersokelse 230
8. Langmyhr FJ, Sveen S (1965) Anal Chim Acta 32:1
9. Bernas B (1968) Anal Chem 40:1682
10. Dolezal J, Lenz J, Sulcek Z (1969) Anal Chim Acta 47:517
11. Kotz L, Kaiser G, Tschöpel P, Tölg G (1972) Fresenius Z Anal Chem 260:207
12. Sulcek Z, Povondra P, Dolezal J (1977) CRC Crit Rev Anal Chem 6:255
13. Tschöpel P, Kotz L, Schulz W, Veber M, Tölg G (1980) Fresenius Z Anal Chem 302:1
14. Kahane E (1937) Fresenius Z Anal Chem 111:14
15. Sichere Chemiearbeit, Mittbl Berufsgen Chem Ind (1983) 35:85
16. Analytical Methods Committee (1959) Analyst 84:214
17. Würfels M (1988) Dissertation Ruhr University Bochum, Germany

18. Kaiser H (1966) Fresenius Z Anal Chem 216:80
19. Jackwerth E, Gomiscek S (1984) Pure Appl Chem 56:479
20. Göller O, Steyrer H, Doktor KJ (1984) Druckbehälterverordnung. Verlag C Heymanns, Cologne, Germany
21. Göller O, Steyrer H, Doktor KJ, Druckbehälterverordnung (1985) Technische Regeln Druckbehälter, Verlag C Heymanns, Cologne, Germany
22. Neumüller OA (1977) Römpps Chemie Lexikon. Frankh'sche Verlagshandlung Stuttgart, Germany
23. Zief M, Mitchell JW (1976) Contamination Control in Trace Element Analysis. Wiley, New York
24. Angst-Pfister, Technische Daten, Produktinformation BE 15387, 11/75 Zürich
25. Kotz L, Henze G, Kaiser G, Pahlke S, Veber M, Tölg G (1979) Talanta 26:681
26. Kaiser G, Götz D, Tölg G, Knapp G, Maichin B, Spitzy H (1978) Fresenius Z Anal Chem 291:278
27. Knapp G (1984) Fresenius Z Anal Chem 317:213
28. Mitchell JW (1973) Anal Chem 45:492A
29. Plastic Hoechst Hostaflon Company Report Hoechst AG
30. Würfels M (3/1989) Labo 7
31. Stoeppler M, Backhaus F (1978) Fresenius Z Anal Chem 291:116
32. Heinrichs H (1989) Laborpraxis 1140
33. Stoeppler M, Müller KP, Backhaus F (1979) Fresenius Z Anal Chem 297:107
34. Sunderman jr FW, Wacinski ET (1974) Ann Clin Lab Sci 4:299
35. Tyler LJ (1973) Chem Eng News 32:20
36. Eustermann K, Seifert D (1977) Fresenius Z Anal Chem 285:253
37. Würfels M, Jackwerth E (1985) Fresenius Z Anal Chem 322:354
38. Tölg G (1975) Pure Appl Chem 44:645
39. Adrian WJ (1971) At Absorpt Newsl 10:96
40. Sinko I, Gomiscek S (1972) Mikrochim Acta 163
41. Paus PE (1972) At Absorpt Newsl 11:129
42. Bernas B (1970) At Absorpt Newsl 9:52
43. Holak W, Krinitz B, Williams JC (1972) J Assoc Off Anal Chem 55:741
44. Holak W (1974) Amer Lab 6:10
45. Holak W (1975) J Assoc Off Anal Chem 58:777
46. Holak W (1980) J Assoc Off Anal Chem 63:485
47. Nelson G, Smith DL (1972) Proc Soc Anal Chem 1968
48. Franco V, Holak W (1975) J Assoc Off Anal Chem 58:293
49. Hartstein AM, Freedman RW, Platter DW (1973) Anal Chem 45:611
50. Würfels M, Jackwerth E, Stoeppler M (1989) Anal Chim Acta 226:1
51. Würfels M, Jackwerth E, Stoeppler M (1988) Fresenius Z Anal Chem 330:160
52. Hendel Y (1973) Analyst 98:450
53. Gomiscek S, Hudnik V, Veber V (1977) Development in Toxicology and Environmental Science Vol 1, Clinical Chem and Chem Toxicology of Metals, Elsevier, Amsterdam
54. Langmyhr FJ, Paus PE (1968) Anal Chim Acta 43:397
55. Langmyhr FJ, Paus PE (1969) Anal Chim Acta 47:371
56. Omang SII, Paus PE (1971) Anal Chim Acta 56:393
57. Buckley DE, Cranston RE (1971) Chem Geol 7:273
58. Grobenski Z (1973) Perkin Elmer Analysentechn Ber 31E
59. Langmyhr FJ, Paus PE (1968) Anal Chim Acta 43:508
60. Langmyhr FJ, Paus PE (1969) Anal Chim Acta 45:157
61. Langmyhr FJ, Paus PE (1969) Anal Chim Acta 43:173
62. Langmyhr FJ, Paus PE (1979) Anal Chim Acta 50:515
63. Hertz J, Pani R (1987) Fresenius Z Anal Chem 328:487
64. Würfels M, Jackwerth E, Stoeppler M (1989) Anal Chim Acta 226:17
65. Pratt KW, Kingston HM, MacCrehan WA, Koch WF (1988) Anal Chem 60:2024
66. Hasse S, Schramel P (1983) Microchim Acta Part III 449

67. Oehme M, Lund W (1979) Fresenius Z Anal Chem 298:260
68. Baumgardt B (1982) MSc Thesis, Ruhr University, Bochum, Germany
69. Adeloju SB, Bond AM, Briggs MH (1984) Anal Chem 56:2397
70. Würfels M, Jackwerth E, Stoeppler M (1989) Anal Chim Acta 226:31
71. Kinard JT (1977) Anal Lett 10:1147
72. Griepink B (1984) Fresenius Z Anal Chem 317:210
73. Knapp G (1984) Fresenius Z Anal Chem 317:213

Chapter 13
Microwave-Assisted Decomposition

JUTTA BEGEROW and LOTHAR DUNEMANN

13.1
Introduction

The pursuit of fast and reliable digestion techniques that are safe and easy to operate in the daily routine work of an analytical laboratory is one of the basic activities of elemental analysis today. Without such techniques modern analytical methods such as atomic absorption spectrometry (AAS), plasma atomic emission spectrometry (ICP-AES, DCP-AES) and plasma mass spectrometry (ICP-MS) would not give correct results in the determination of elements in complex samples, because the solubilisation of analytes and removal of interfering molecules are essential for correct analyses.

Microwave energy as a heat source for ashing procedures has become a reasonable link between the time-consuming sample decomposition techniques and the fast determination methods. Only recently, this link has led to completely automated on-line digestion/determination techniques.

In 1975 the first paper on microwave-assisted decomposition was published by the group of Koirtyohann [1]. The first compendium on microwave sample preparation was edited by Kingston and Jassie in 1988 [2]. In the early nineties the introduction of high temperature/pressure devices gave a further impulse to the development of microwave-assisted decomposition [3].

13.1.1
Fundamentals of Microwave Systems

The earliest attempts at microwave-assisted decomposition were performed using home appliance microwave ovens. This was necessary because commercial devices were not available at that time.

Modern commercial instrumentation for microwave-assisted decomposition (Fig. 13.1) is equipped with a magnetron operating at a frequency of 2.45 GHz. Microwaves cannot rupture molecular bonds directly because the corresponding energy is too low to excite electronic or vibrational states. Rotational excitation of dipoles and molecular motion associated with the migration of ions are the only processes that are observed in this microwave field [4, 5]. For this reason the designation "microwave-*assisted* decomposition" is recommended.

Microwave radiation, generated by a magnetron, is directed via a waveguide into the oven cavity. The power output of the "cycled" magnetron is controlled by the contribution of the duty cycles. If the duty cycle is 50% of the constant time basis

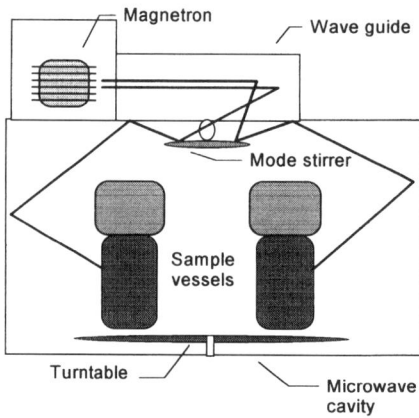

Fig. 13.1. Schematic illustration of a microwave-assisted digestion apparatus

of the whole cycle, then the actual power output is also 50% of the maximum power output. The power output of a microwave oven is between 600 and 1200 W. For even distribution of the radiation a "mode stirrer", a rotating antenna or a turntable are provided. This is necessary to prevent the development of standing waves and uncontrolled local overheating problems.

Microwave energy which is introduced as radiation into a liquid medium is converted into heat by oscillating alignment and subsequent relaxation of dipoles (about 5×10^9 times per second at 2.45 GHz) as well as by the collision of ions with solute molecules during their migration in the field. The contribution of ionic conduction is enhanced if movable (mobile) ions are present in the sample solution, which can be demonstrated by the addition of sodium chloride. Both dissipation processes cause the rapid heating of the samples.

Different absorption properties require a distinction between two catagories of matter. (1) Matter that cannot absorb microwaves (conductors and insulators): Metals reflect microwave radiation whereas quartz glass or fluoro-compounds such as PTFE are transparent for microwaves. Neither of them will heat up because there is no absorption of microwave energy. (2) Matter with absorption properties (dielectric materials): An aqueous digestion solution (e.g. a mixture of a sample with a strong mineral acid) behaves as an absorptive material that will be heated rapidly by the dissipation of microwave radiation.

The penetration depth of microwaves into an absorptive material depends on its dissipation factor (tan δ). Materials with a high dissipation factor (salt solutions, water) have a low penetration depth, those with a low dissipation factor are nearly transparent. The penetration depth of microwaves in water is a few centimeters.

13.1.2
Fields of Application

Since the mid-eighties microwave-assisted decomposition techniques have been performed by an increasing number of laboratories for trace analysis in geological or biological samples as well as for material analysis and purity control analysis [6–9].

Today there is a lot of experience with this fast sample digestion technique. We are sure that microwave-assisted digestion is already a precious tool for reproducible and correct trace-element determinations in biological, environmental and material sciences. The cooperation between manufacturers and users has led to systems which are as reliable as those that use conventional convective heating.

13.2
Microwave-Assisted Decomposition Apparatus

13.2.1
Safety Precautions

Pressure digestion techniques must always be carried out with exceptional care and attention. The ashing of organic matrices can cause the pressure to rise to very high levels within a short period of time, and the reaction may become uncontrolled. The risk of severe damage because of a ruptured pressure vessel cannot be overemphasized. The digestion of an unknown matrix requires special care; it is important to begin with a small amount of sample, whose mass should not exceed 100 mg.

Additional safety precautions apply to the use of microwave ovens. The equipment should be operate in strict accordance with the instructions to ensure that not more than the allowed amount of microwave radiation will escape. Leak tests should be conducted on a regular basis by a trained technician. Vessels or other objects made of metal should never be placed in the oven chamber. The oven must never be operated when it is empty; a microwave absorbing substance should always be present (water, sample solution, etc.), since the magnetron might be destroyed by reflected radiation. A system equipped with a pressure sensor is valuable in this respect, and temperature control can also be advantageous, especially for method development [10].

13.2.2
Microwave-Assisted Digestion Techniques

Pressure digestion allows the sample to be essentially isolated from the laboratory atmosphere, thus minimizing both the loss of volatile analytes and contamination from particles in the air. Moreover it has the advantage that the digestion takes place at higher temperatures compared to digestion at atmospheric pressure, because of the boiling-point-elevation effect. The pressure itself can be regarded as nothing more than an unavoidable side effect. The high temperature has a strong influence on the oxidizing power of the digesting acids. That is why we have to distinguish between low-pressure (up to about 10 bar) and high-pressure systems (up to about 70 bar or higher). The latter systems are particularly suitable for trace analysis in thermally resistent matrices. They also leads to a very low residual carbon content in the digestion solution, thus allowing polarographic or voltammetric analysis to be performed.

A disadvantage of pressurized systems are their low degree of automation. Here non-pressurized systems offer some special features such as automated serial analyses. Microwave-assisted techniques are especially valuable because the fast heating of the sample allows on-line analyses in combination with AAS, ICP-AES or ICP-MS.

13.2.2.1
Low-Pressure Systems with Home Appliance Microwave Ovens

Lightly contaminated water, many geological samples and food with low fat content can be completely decomposed using home appliance microwave ovens with low pressure vessels [11, 12]. In this case it is necessary to follow the above-mentioned safety precautions. The amount of sample should be low with respect to the carbon content. Some examples of microwave-assisted digestions are given in Table 13.1.

Digestion with a single acid, e.g. nitric acid, is not always successful with such simple systems, so the addition of hydrochloric acid or hydrogen peroxide will be necessary. If silicate-bearing materials are to be digested, the addition of hydrofluoric acid is unavoidable. The maximum attainable pressure seldom exceeds 10 bar without destruction of rupture disks and other safety features.

Another severe disadvantage of this simple approach is the frequent need for multi-step digestion, in which a preliminary digestion is followed by further treatment, often with hydrogen peroxide [10]. The distribution of microwaves in most home appliance microwave ovens is not as homogeneous as it should be for reproducible digestions. This is due to the inhomogeneous microwave field in such ovens, even in those with turntables and antenna. As a rule, the power should be kept low and the digestion period lengthened.

The costs of such simple digestion equipment are very low, but here a word of warning is necessary: home appliance microwave ovens with pressure vessels should be used with special care because these ovens are not equipped with additional safety features.

Table 13.1. Comprehensive description of home appliance microwave ovens with low pressure vessels [8, 9]

Biological samples (e.g. fat-rich plants or food)	
Sample size	100–200 mg
Addition of acid	2 ml HNO_3
Ashing steps	3 min/280 W, 3 /420 W, 2 min/560 W; cooling phase between treatments: 1 min
In case of uncomplete ashing	2 /560 W after addition of 1 ml H_2O_2
Geological samples (e.g. soils and ore slags)	
Sample size	100 mg
Addition of acid	2 ml HNO_3 + 2 ml HCl + 2 ml HF
Ashing steps	3 min/140 W, 3 min/280 W, 3 min/420 W, 1 min 560 W; cooling phase between treatments: 1 min

13.2.2.2
Commercial Low-Pressure Microwave Systems

The introduction of improved completely commercial microwave systems for sample digestion in the late eighties [13] has led to a further optimiztion of fast sample preparation. Commercial systems offer additional safety features and improved facilities for pressure and/or temperature control.

For the reliable digestion of complex matrices and for safety reasons, these systems are to be preferred to home appliance microwave ovens, although they are more expensive. The distribution of microwave radiation inside the oven cavity is more homogeneous, especially because the magnetron is subject to shorter switching cycles. These features are of special value for all routine work.

It is important for the reproducibilty and reliability of microwave-assisted decomposition to know what kind of safety mechanisms are available in the case of too a high pressure. Severe losses of analytes can occur if sample components are blown off. It is essential to be aware that such an event has to be recorded reliably. This is not a problem at all in case of a rupture disk, because the destruction is irreversible and can be easily recognized. In the case of a reversible mechanism, this blow off may not be notized, with the result that the operator cannot give the reason for a false result. On the other hand, reversible mechanisms can avoid bigger losses for they reclose the system immediately after release of the overpressure. The user has to decide whether a reversible or an irreversible mechanism is the better alternative for his particular application. For method development, the irreversible rupture disk is preferable.

Low-pressure systems allow decomposition temperatures of about 180 °C, a level which is suitable for the digestion of most geological samples and biological samples with a low fat content. For many matrices, such temperatures are not sufficient to guarantee the complete ashing of thermoresistent sample components. This is especially problematic for matrix-sensitive determination methods.

13.2.2.3
High-Pressure Microwave Systems

High-pressure/high-temperature equipment is designed to support digestion up to a pressure, of 70 bar or more. Substantially higher temperatures (up to above 250 °C) can be reached, resulting in more thorough digestion, even of difficult matrices [10, 14, 15]. Temperature and digestion time determine the effectiveness of a decomposition. For organic samples, the residual carbon content serves as a useful measure for this effectiveness. The advantage of high-pressure digestion is that nearly every sample matrix can be ashed completely by a single-stage process, even in the case of high-fat matrices (Table 13.2).

Electrochemical determination methods such as voltammetry or polarography are especially sensitive to residual organic compounds, because they may influence the electrode potential.

Besides the temperature, the time period of the process is also important for the ashing result. The slogan "microwave ashing in 60 seconds" should be regarded as

Table 13.2. Comprehensive description of the commercial microwave system PDM with high pressure vessels [13]

Biological samples (e.g. fat-rich plants or food)	
Sample size	100 mg
Addition of acid	2 ml HNO_3
Ashing conditions	10 min/500 W
Cooling	1 min
Biological samples (e.g. animal or human tissues)	
Sample size	150–200 mg
Addition of acid	2 ml HNO_3
Ashing conditions	10 min/500 W
Cooling phase	10 min

scientific humbug, because this is simply the time needed for heating a sample/acid solution to the desired temperture (180 °C in low-pressure microwave systems, 250 °C in high-pressure systems). For the complete decomposition of a matrix, even a simple one, it is self-evident that this temperature has to be held for at least 10 minutes.

13.2.2.4
Non-Pressurized Microwave Systems

This relatively inexpensive technique may become a standard technique for clearly defined problems in trace analysis. It is designed for routine use and can easily be automated. All relevant parameters, such as as time, temperature and addition of reagents lend themselves to straightforward control.

Non-pressurized microwave systems are limited by a low maximum ashing temperature, that cannot exceed the ambient-pressure boiling point of the acid (or the acid mixture). As stated earlier, the oxidizing power of nitric acid is insufficient at such low temperatures (ca. 122 °C). One possible remedy is the addition of sulfuric acid, which significantly increases the boiling point of the sample solution. It always should be kept in mind that sulfuric acid forms stable compounds, e.g. with lead in graphite furnace AAS, so that the practicability depends on the matrix and the analyte [12]. High-fat and high-protein samples are mostly not subject to complete ashing at ambient pressure. Other disadvantages are the danger of contamination by the air in the laboratory and the danger of trace losses, especially if mercury is the analyte. Losses of organo-metallic compounds of arsenic, antimony and tin may also occur if such ashing is performed without precautions.

On the other hand, ambient pressure systems are the best option with regard to the safety of personnel, because no overpressure can occur. Moreover, non-pressurized microwave digestion is suitable for on-line decompositions in continuous-flow systems [16].

13.2.2.5
Dry Ashing in a Microwave Oven

For geological and metallurgical samples, dry ashing in a microwave oven is a sensible alternative to conventional dry ashing in a muffle furnace. The sample is positioned in a special block fabricated from a material that is highly absorbent for microwave radiation (e.g. silicon carbide) and is surrounded by quartz insulation. This block can be heated to about 1000 °C within 2 min at maximum power of the microwave oven. The advantages compared to conventional dry ashing techniques are striking: It is time saving, the energy consumption is low and it spares the laboratory personnel from exposure to intense heat when inserting and removing samples [2, 4].

13.3
Comparison of Different Microwave-Assisted Digestion Systems

The digestion of samples with high content of proteins and fatty compounds is especially problematic, because during the ashing procedure there thermo-resistent reaction products may be formed. Ashing with low maximum temperatures, as in non-pressurized or low pressure systems, may be uncomplete. This is even more problematic with microwave-assisted digestion, because the ashing time is particular short (typically 10 min). The addition of perchloric acid or hydrogen peroxide should only be done after taking precautions: the formation of reactive radicals may lead to an explosive reaction.

Reviews on commercially available microwave-assisted digestion systems are given in [2, 13]. For a comparison of protein- and fat-rich samples the following digestion systems were tested.

13.3.1
Microwave-Assisted Digestion Systems

Low-pressure system with home appliance microwave ovens: Digestion vessels DAP 50 (50 ml) from Berghof, Eningen, or digestion vessels Parr-Instruments No 4781 (23 ml) from Kürner, Rosenheim in a domestic microwave oven (with and without turntable, 650 and 700 W, respectively).

Commercial low-pressure systems: MLS 1200 (Büchi, Göppingen, also available with high-pressure vessels) and MDS-81 (CEM, Moers).

Commercial high-pressure system: PMD ("Pressurized microwave digestion", Kürner, Rosenheim).

Non-pressurized digestion system: Microdigest 300M (200 W, Prolabo, sales distribution in Germany: Pabisch, München) with Kjeldahl vessels (30 ml).

13.3.2
Conventional Digestion Procedures

Convective thermal low-pressure digestion: Refined steel autoclaves with PTFE inserts in a heating block with control of time and temperature (Berghof, Eningen).

Convective thermal high-pressure digestion: HPA ("High-pressure asher", Kürner, Rosenheim).

Cold-plasma ashing: Plasma-Processor 200-G (Technics Plasma) with petri dishes.

Experimental details are given in [12] for non-pressurized and low-pressure microwave-assisted systems and in [17–19] for conventional systems.

The function of the microwave-assisted high-pressure system PMD is given here in a very comprehensive way: The PMD consists of a microwave oven with a venting and cooling facility. The maximum power output is 750 W, power and digestion time can be adjusted stepwise (power: 12 steps; time: 0–60 min). Two pressure containers can be handled simultaneously in the oven. Therefore the sample together with a defined amount of concentrated acid is given into the inner sample vessel made of quartz glass or PFA (perfluoralkoxy), which is then closed with a titanium stopper and a rupture disk and positioned in the pressure container made of a microwave-transparent high-performance polymer. Finally the pressure container is closed with a screw top and put into an additional protective case. During the decomposition phase, an additional safety facility on a light sensor basis ensures rapid cut-off of the microwave power if the reaction pressure is too high e.g. 80 bar.

13.4
Criteria for the Evaluation of Microwave Digestion Systems

The results of a microwave-assisted digestion procedure can be best evaluated by the following criteria: completeness of matrix destruction, reproducibility with respect to the analyte determination, freedom from trace losses and contamination, ease of handling, and expenditure of time. Whether or not a microwave system is suitable for a special trace analytical problem depends on these criteria.

Residual carbon content in the digestion solution: Although some methods of determination do tolerate certain amounts of residual carbon (e.g. in atomic spectrometry), one must be aware of severe faults if there is no or only inadequate correction of molecular interferences. For most electrochemical determination techniques (e.g. voltammetry), complete decomposition of organic substances is necessary because such compounds are electrochemically active and show a strong tendency to interfere with the analyte signal [15]. Taking milk as an example, the interference of thermo-resistant compounds in electrochemical determinations has been investigated [10].

In case of AAS or ICP-MS, a digestion solution that is colourless and free of particles will suffice. Although all the techniques described here satisfy this requirement, only the high-pressure techniques are suitable for electrochemical determinations. Complete destruction of fat- and protein-rich samples (> 99%) can only be guaranteed by using the convective high-pressure technique. The microwave-assisted high-pressure devices are not as effective, possibly because of the short period of time during which samples are exposed to high enough temperatures. According to information from the distributor, the residual carbon content in digestion solutions of the PMD is about 94% for beef liver and about 98% for milk powder.

Reproducibility with respect to the analyte determination, freedom from trace losses and contamination: Microwave-assisted ashing of fat-rich samples in open vessels is critical with respect to losses of volatile analytes, despite the use of a reflux condenser. The reason may be overheating phenomena occur in the ashing solution as a result of the rapid heating process. On the other hand this technique is suitable for ashing low-fat matrices: Only for mercury were significant losses found, for selenium there was good recovery (97.5% in the case of inorganic Se compounds, see Table 13.3). This experiment was not performed with organic Se compounds, possibly because the higher volatility of organic compounds leads to higher losses.

For the pressurized microwave techniques, no higher losses were observed than for the comparable convective systems.

Trace contaminations in ashing techniques are due to the reagents used, the ambient air and the constitution of the vessel walls that come into contact with the digestion solution. Memory effects are often lower in microwave-assisted techniques than in conventional techniques, which may be due to the shorter contact time in the microwave vessels.

In Table 13.4 the reproducibilities of at least five repeated digestions for the different ashing techniques are compared. The non-pressurized microwave technique shows the lowest standard deviation, which may be explained by the low memory effects that were observed in this case.

Expenditure of time: The microwave-assisted techniques allow samples to decompose rapidly (Table 13.5a). To estimate the time taken for digestion, not only was the ashing phase considered but also the time necessary to prepare the sample (aliquotation, addition of acid), the cooling phase, the transfer of the digestion solution to the determination system and the period for cleaning the vessels and all other utilities. According to Table 13.5b, pressurized microwave digestion is the fastest of all the techniques tested, followed by non-pressurized microwave ashing. The conventional high pressure technique is 3 or 4 times slower when the simul-

Table 13.3. Recovery of selenium and mercury in soya flour after non-pressurized microwave-assisted ashing. The two elements had been introduced beforehand by adding 20 µg/l each of Se (inorg.) and Hg (inorg.) to the samples (n = 9)

Analyte	Recovery (%)
Se	97.5 ± 6.8
Hg	26.0 ± 12.5

Table 13.4a. AAS data obtained after ashing of foods with medium fat content (in mg/kg)

Ashing technique	Zn	Fe	Cu	Ni
Soya flour				
MW non-pressurized	33.4 ± 2.7	48.3 ± 3.1	9.9 ± 0.6	7.1 ± 0.5
MW low pressure	38.2 ± 1.9	58.3 ± 6.2	11.7 ± 1.8	9.4 ± 1.4
Cold plasma ashing	36.3 ± 2.0	56.9 ± 4.6	11.3 ± 1.1	9.3 ± 2.8
Convent. low pressure	41.6 ± 6.0	60.1 ± 2.4	11.8 ± 0.8	10.8 ± 1.0
Linseed				
MW non-pressurized	29.1 ± 1.8	36.4 ± 2.6	5.7 ± 0.6	3.0 ± 1.0
MW low pressure	39.5 ± 3.1	56.3 ± 8.9	10.0 ± 1.3	5.5 ± 1.4
Cold plasma ashing	35.8 ± 4.7	53.3 ± 2.5	8.6 ± 1.1	4.5 ± 1.8
Convent. low pressure	33.9 ± 2.2	62.0 ± 7.5	8.2 ± 1.1	5.5 ± 1.5

Table 13.4b. AAS data obtain after ashing of foods with high fat content (in mg/kg)

Ashing technique	Zn	Fe	Cu	Ni
Peanut butter				
MW non-pressurized	21.1 ± 0.8	20.9 ± 1.7	3.2 ± 0.3	2.3 ± 0.5
MW low pressure	25.8 ± 1.2	25.6 ± 1.5	5.4 ± 0.6	3.0 ± 0.8
Cold plasma ashing	23.3 ± 2.0	23.1 ± 1.5	5.2 ± 0.5	2.3 ± 0.4
Convent. low pressure	25.1 ± 2.0	25.9 ± 4.6	5.1 ± 0.7	3.8 ± 1.6

Table 13.5a. Time (in minutes) required for high pressure ashing with microwave and conventional convective heating (two samples simultaneously)

	Microwave assisted high pressure ashing	Conventional high pressure ashing
Preparation steps	10	10
Ashing phase	10	90–150
Cooling phase	10	20
Transfer step	5–10	5–10
Cleaning step	10–20	10–20
Sum	45–60 min	135–210 min

Table 13.5b. Comparison of the time required for the ashing of two samples (min)

Matrix	Microwave assisted ashing		Conventional assisted ashing	
	high pressure	non-pressure	high pressure	low pressure
Cereals	45	120	140	270
Sun flower oil	50	120	160	270

taneous digestion of two samples is performed. Much slower is the cold-plasma ashing technique (10–20 h), which in many laboratories serves as a reference method or as a method for the digestion of extremely resistant matrices but is not involved in daily routine analysis.

The expenditure of time is a very important criterion for all laboratories that have to deal with "urgent" sample analyses. The time needed for the digestion of a sample can is this case be regarded as the "bottleneck" of the analysis.

Ease of handling: The handling of commercially available microwave-assisted digestion systems is easy to learn. The microwave ovens are safe with respect to microwave radiation and high pressure. If the above-mentioned safety precautions are strictly followed, the risk for laboratory personnel should be very low. The non-pressurized techniques can be easily automated, and on-line digestions of simple matrices are another important option of this technique [16].

13.5
Conclusions

Sample preparation, in particular the digestion step is increasingly becoming the "bottleneck" of analysis, because the actual determination methods are very fast when compared with traditional methods of decomposition. Comparisons of the time required for different decomposition techniques are only valid if the whole digestion procedure is taken into account, i.e. not simply the ashing phase itself, which lasts about 10 minutes. In particular the long cooling phase, that is nearly as long as the ashing phase (the digestion solution must be cooled down to room temperature before opening the vessel), contributes to the overall processing time of microwave-assisted ashing.

Microwave-assisted techniques are 3 or 4 times faster than most of the conventional techniques. Nevertheless, serious manufacturers and distributors of microwave systems should avoid the slogan "microwave ashing in 60 seconds" as stated above, and it should be kept in mind that microwaves cannot rupture molecular bonds. The reduction in time is rather a result of the rapid heating process, because microwaves can penetrate the vessel walls and can therefore heat the sample/acid solution directly.

The rapid heating process and the short ashing cycle produce other advantages for trace analysis: The materials involved (vessels, screw caps, seals) are subject to less stress than in convective heating processes. There is thus less tendency for porous surfaces to form. Surfaces (e.g. vessel walls) have a tremendously high surface area that may lead to high adsorption and desorption rates for trace metals.

Despite the important advantage of reducing sample preparation time, microwave-assisted digestion is not yet the method of choice if complete digestion is necessary, which is the case for most electrochemical determination methods. Because the ashing phase of microwave systems is short, complete decomposition of thermo-resistant matrix components is not always guaranteed. In this area of application, conventional high pressure ashing is preferable.

13.6
References

1. Abu-Samra A, Morris JS, Koirtyohann SR (1975) Anal Chem 47:475
2. Kingston HM, Jassie LB (eds) (1988) Introduction to Microware Sample Preparation. American Chemical Society, Washington DC
3. Dunemann L, Begerow J, Bucholski A (1994) Sample Preparation for Trace Analysis. In: Ullmann's Encyclopedia of industrial chemistry. Verlag Chemie, Weinheim, Vol. B 5:65
4. Neas ED, Collins MJ (1988) In: Kingston HM, Jassie LB (eds) Introduction to Microwave Sample Preparation. American Chemical Society, Washington DC, p 7
5. Nadkarni RA (1984) Anal Chem 56:2233
6. Sulcek Z, Povondra P (1989) Methods of decomposition in inorganic analysis. CRC Press, Boca Raton (Florida)
7. Aysola P, Anderson P, Langford CH (1987) Anal Chem 59:1582
8. White RT, Douthit GE (1985) J Assoc Off Anal Chem 68:766
9. Xu L-Q, Shen W-X (1988) Fresenius Z Anal Chem 332:45
10. Reid HJ, Greenfield S, Edmonds TE (1995) Analyst 120:1543
11. Dunemann L (1989) In: Welz B (ed) 5. Colloquium Atomspektrometrische Spurenanalytik. Bodenseewerk Perkin-Elmer, Überlingen, 593
12. Dunemann L, Meierling M (1992) Fresenius J Anal Chem 342:714
13. Dunemann L (1991) Nachr Chem Tech Lab 39(10):M1
14. Würfels M, Jackwerth E, Stoeppler M (1987) Fresenius Z Anal Chem 329:459
15. Hertz J, Pani R (1987) Fresenius Z Anal Chem 328:487
16. Karanassios V, Li FH, Liu B, Salin DS (1991) J Anal Atom Spectrom 6:457
17. Kotz LG, Kaiser L, Tschöpel L, Tölg G (1972) Fresenius Z Anal Chem 260:207
18. Schwedt G, Dunemann L (1990) Labor Praxis, Juni 1990, 476
19. Schramel P, Haase S, Knapp G (1987) Fresenius Z Anal Chem 326:142

Chapter 14
Decomposition Methods for the Electrochemical Determination of Elements

Peter Ostapczuk

Index of Abbreviations

DPASV	Differential pulse anodic stripping voltammetry
CPE	Carbon paste electrode
GCE	Glassy carbon electrode
DC	Direct current
CME	Chemically modified electrodes
ADPV	Adsorptive differential pulse voltammetry
HMDE	Hanging mercury drop electrode
DPP	Differential pulse polarography
DME	Dropping mercury electrode
iDC	Stripping direct current
AuE	Gold electrode
AuFE	Gold fibre electrode
RAuE	Rotating gold electrode
PSA	Potentiometric stripping analysis
MFE	Mercury film electrode
GCRDE	Glassy carbon rototating disc electrode
AC1	Alternating current voltammetry
iAC	Stripping alternating current voltammetry
RMFE	Rotating mercury film electrode
Fl. Sys.	Flow system
CFE	Carbon fibre electrode
SMDE	Static mercury drop eletrode

14.1
Introduction

Stripping analysis is an extremely sensitive electrochemical technique for measuring trace metals [1, 2]. Its remarkable sensitivity is attributed to the combination of an effective preconcentration step with advanced measurement procedures that generates an extremely favourable signal-to-background ratio. Essentially, stripping analysis is a two-step technique. The first, or deposition step, involves the electrolytic deposition of a small portion of the metal ions in the solution into the mercury electrode to preconcentrate the metals. This is followed by the stripping step (the measurement step), which involves the dissolution (stripping) of the deposit.

Anodic stripping voltammetry (ASV) is the most widely used form of stripping analysis. In this case, the metals are preconcentrated by electrodeposition into a small-volume mercury electrode (a thin mercury film [MFE] or a hanging mercury drop [HMDE]). The preconcentration is done by cathodic deposition during a controlled time and at a fixed potential. The deposition potential is usually 0.3–0.5 V more negative than $E°$ for the least easily reduced metal ion (e.g. zinc, cadmium, lead, and copper) to be determined. The metal ions reach the mercury electrode by diffusion and convection. The convective transport is achieved by electrode rotation (for MFE) or solution stirring (for MFE or HMDE). The duration of the deposition step is selected according to the concentration of the metal ion in question; it generally varies from 0.5 min for a 10^{-7} M solution to 10 min (for HMDE) or 30 min (for MFE) for a 10^{-11} M solution. After preselecting the time of deposition, the forced convection is stopped, and the potential is scanned anodically, linearly, or in a more sensitive potential-time (pulse) waveform that discriminates against the charging background current (usually square-wave [SW-ASV] or differential-pulse [DP-ASV] ramps). During this anodic scan, the amalgamated metals are reoxidized, and stripped out of the electrode. The voltammetric peaks (potential-time curve) reflect the time-dependent concentration gradient of the metal in the mercury electrode during the potential scan. Peak potentials serve to identify the metals in the sample and the peak current is proportional to the concentration.

Cathodic stripping voltammetry (CSV) is the mirror image of ASV. It involves anodic deposition of the analyte, followed by stripping during a negative potential scan. The resulting reduction peak current provides the desired quantitative information. CSV is used to measure a wide range of organic and inorganic compounds capable of forming insoluble salts with mercury (e.g., halide ions) or intermetallic compounds in mercury (Se and As with Cu).

Potentiometric stripping analysis (PSA), developed largely by Jagner [3], differs from ASV in the method used for stripping the amalgamated metals. In this case, the potentiostatic control is disconnected and the preconcentrated metals are reoxidized by an oxidizing agent (e.g., oxygen, mercury ions) which is present in the solution. During the oxidation step, the change in the working electrode potential vs. time is recorded. A sharp potential step accompanies the depletion of each metal from the electrode. The time needed for the oxidation (transition time) of a given metal is a quantitative measure of the metal concentration in the sample. Modern PSA instruments use microcomputers to register fast stripping events (as the time can be monitored with extremely high resolution by the computer clock) and to convert the wave-shaped response to a more convenient peak over a flat baseline. The use of nondeaerated samples represent an important advantage of PSA. In addition, this method can accommodate electrodes of any size, thus eliminating the need for amplification when microelectrodes are used.

Adsorptive stripping analysis enhances the scope of stripping measurements for numerous trace elements [4, 5]. The relatively new strategy involves the formation, adsorptive accumulation, and reduction of a surface-active complex of the metal. Both voltammetric and potentiometric stripping schemes, with a negative potential scan or constant cathodic current, respectively, can be employed for measuring the adsorbed complex. Short adsorption times (0.5–5 min) result in a very effective interfacial accumulation. The reduction step is also very efficient as all of the collected complex is reduced. The limitation is the adsorption of one monolayer with the result that the calibration curves display nonlinearity at high concentrations.

14.2
Determination without Sample Decomposition

By potentiometric stripping analysis (PSA) it is possible to determine some elements in liquid samples after simple sample preparation. Due to the chemical equilibria in the solution and due to the limited potential range, this method is limited to the determination of Cd, Pb and Tl. Appropriate sample preparation is the most important step in the entire procedure. The liquid samples must be acidified to liberate all the ions from their organic complexing agents. Depending on the matrix, a pH value between 3 and 1 is selected for ion liberation. Lead can be determined without any trouble in whole blood, serum, urine, wine, beer, milk, juices, waste, water, natural waters, sea water and all other liquid samples at pH 1. The determination of cadmium and thallium in these samples is more complicated because of a significantly lower concentration level, and the determination procedure must be developed for each kind of sample. In some cases it is also possible to determine these elements in solid samples, provided they can be dissolved in water or an appropriate supporting electrolyte (e.g. sugar or certain medicaments).

Anodic stripping voltammetry, unlike PSA, is very sensitive to the presence of organic compounds in the sample solution. For this reason it is necessery to separate the big organic molecules from the electrode either by covering the electrode surface with a special chemical membrane or by changing the supporting electrolyte after the deposition step. The latter technique is onyl possible in flow systems, which are commercially available. The future development will demonstrate whether this technique can be adopted for routine analysis.

14.3
Oxidative UV-Digestion

Dissolved organic matter (DOM) is a significant component of many natural water matrices, such as river, lake, ground, estuarine and coastal water. In municipal and special industrial waste water and water of sewage treatment plants, the DOM concentration is very high.

As a substance-specific method, voltammetry offers a powerful means of investigating and studying the speciation of heavy metals dissolved in natural waters. On the other hand, certain dissolved metal species, particularly rather stable organic complexes, are hardly, or not at all, accessible to voltammetric determination. Therefore, the determination of the complete content of dissolved heavy metals in natural waters requires appropriate sample pre-treatment to transform all the dissolved heavy metals into chemical forms well accessible to voltammetric determination [6, 9].

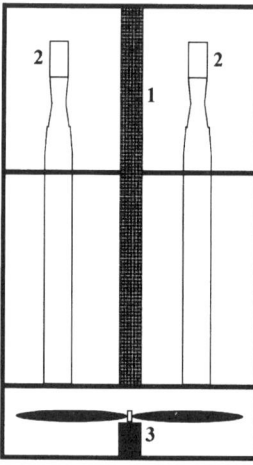

Fig. 14.1. UV irradiation system: (*1*) UV-lamp; (*2*) sample vessels made from quartz; (*3*) cooling system

The DIN standard 38 406, Part 16 [7] describes the voltammetric determination of Zn, Cd, Pb, Cu, Tl, Ni and Co in water samples. One of the sample preparation methods used involves digestion by UV irradiation (Fig. 14.1). For this kind of digestion two types of mercury lamps can be used: low-pressure and high-pressure. The low-pressure mercury lamp has a line spectrum with a relatively high fraction of hard UV radiation at 185 nm (8% relative intensity) and the main component at 254 nm (90% relative intensity). The low intensity means that the rate of heating is low and thus the time needed for complete digestion, even after addition of hydrogen peroxide, is several hours. With high-pressure mercury lamps a broad band spectrum between ca. 200 and 435 nm is obtained [8]. Furthermore, these lamps have a significantly higher intensity as well as a greater radiant flux at all wavelengths. An experimental difficulty in the use of high-pressure mercury lamps is that a larger amount of heat is produced, which can lead to high evaporation losses of the samples. Therefore, an efficient sample cooling system is necessary. Using modern UV-digestion systems, even samples strongly loaded with dissolved organic matter (e.g. municipal water) can be digested without any difficulty in only one hour [8].

Table 14.1 presents some examples of elements determined by different electroanalytical methods in liquid samples without sample decomposition and after UV irradiation.

14.4
Mineralization by Oxygen

The use of oxygen for the mineralization of biological materials has a long tradition. There exist a number of methods which are based on oxidation by air, pure oxygen or other oxydizing gases. Heating of biological materials in an electric oven is one of these methods. Dry ashing has only a limited application for electroanal-

Table 14.1. Determination of elements without sample decomposition

Element	Sample	Determination method	References
Ag	River water	LSV/GCE	[14]
Al	Sea water	ADPV/HMDE	[15]
As	Tap water	ADPV/HDME	[16]
	Sea water	PSA/AuFE	[17]
Bi	Sea water	SWV/GCRDE	[18]
Cd	River water	PSA/MFE	[19]
	Blood	PSA/MFE	[20, 21]
	Wine	PSA/MFE	[22]
Co	Natural water	PSA/MFE	[23]
	Sea water	PSA/Flow system	[24]
Cr	Cooling water	ADPV/HMDE	[25]
	Cooling water	DPP/MDE	[26]
	Sea water	ADPV/HMDE	[27]
Cu	Sea water	PSA/Flow system	[24]
	Blood	DPASV/GCA	[25]
	Wine	PSA/MFE	[26, 27]
Fe	Sea water	ADPV/HMDE	[28, 29, 30]
	Natural water	ADPV/MCPE	[31]
	Snow, rain	DPP/MDE	[32]
	Wine	CSV/HMDE	[33]
Mn	Fresh water	ADPV/HMDE	[34]
Mo	Drinking water	ADPV/HMDE	[35]
Ni	Sea water	PSA/Flow system	[23, 24]
Pb	River water	PSA/MFE	[36]
	Blood	DPASV/GCE	[25]
	Blood	PSA/MFE	[37]
	Blood	PSA/AuE	[38]
	Urine	DPASV/GCE	[25]
	Wine	PSA/MFE	[26]
Pd	Sea water	ADPV/HMDE	[39]
	Natural water	ADPV/MCPE	[40]
Pt	Tap water	ADPV/HMDE	[41]
Ru	Tap water	ADPV/HMDE	[41]
Se	Sea water	ADPV/HMDE	[42]
Sn	Natural water	ADPV/HMDE	[43]
	River water	ADPV/HMDE	[44]
Tl	River water	PSA/MFE	[45]
	Urine	DPASV/MFE	[46, 47]
Ti	Tap water	ADPV/HMDE	[48]
U	Sea water	PSA/Flow system	[49]
V	Tap water	ADPV/HMDE	[50]
Zn	Industrial electrolytes	DPASV/HMDE	[51, 52, 53, 54]
	Zn, Cd, Pb	PSA/Flow system	[55]

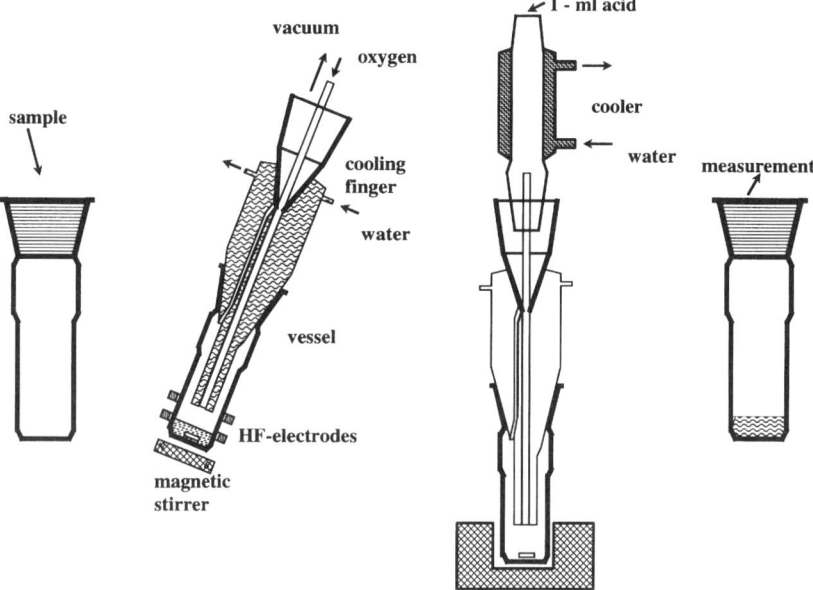

Fig. 14.2. Digestion steps of the Cool Plasma Asher

ytical determination. Due to the loss of cadmium and a contamination risk for copper, nickel and cobalt, this technique can be used only for the determination of lead and zinc in biological materials. The digestion with pure oxygen (Trace-O-Mat) or oxygen plasma increases significantly the number of elements which can be determined by electrochemistry.

The organic sample can be decomposed in the well known manner at low temperature by a high frequency induced oxygen plasma. As opposed to other commercial versions of the plasma asher, in the Cool Plasma Asher (CPA, Fig. 14.2) the sample is contained in a quartz test tube that is equipped with a cooling finger. Volatile elements such as arsenic, selenium and others, that are partially lost in traditional plasma ashers, are retained on the cooling finger during the ashing process; mercury, however, is lost. After the destruction, the liberated elements are dissolved by refluxing in 1–2 ml acid [13].

An advantage of this method is that the elements are present in comparatively high concentrations after digestion because only a small volume of acid used: up to 1 g of sample material can be ashed and the liberated elements collected in only 2 ml of ultra-pure acid by refluxing. For environmental analysis, this method is perfectly suitable for destroying dust-collecting filters made of cellulose or polymers. Even PTFE can be ashed by this method. The inorganic part of the collected dust remains in the quartz test tube for further processing.

Selenium and arsenic can also be determined by electroanalytical techniques in solid and liquid samples after digestion in a Wickbold system (combustion in an oxygen/hydrogen flame, see Chapter 15 in this book).

14.5
Open Wet Digestion

Open wet digestion is a cheap digestion technique and is the one most commonly used. A number of methods are available commercially for this kind of digestion. Only two of them have really found any adaptation for electroanalytical determination. The first one is based on the carbonization of biological materials with sulphuric acid and carbon oxidation by hydrogen peroxide. Using this technique, a lot of environmental samples can be digested for the determination of Zn, Cd, Pb, Tl, Cu, Ni and Co by electroanalytical methods. This mineralization method cannot be used for samples with high calcium concentration (e.g. bones) due to co-precipitation of elements with calcium sulfide.

The second wet open digestion method is based on the oxidation of biological matrices by a mixture of nitric, perchloric and sulfuric acids. The addition of sulphuric acid is not necessary in all types of matrices. This digestion technique is suitable for a number of environmental samples, although the use of the very dangerous perchloric acid is one of its major disadvantages.

Table 14.2. Determination of elements by electroanalytical methods after UV irradiation

Element	Sample	Determination method	References
Al	Polluted water	ADPV/GCE	[56]
Cd	Sea water	DPASV/MFE	[57, 58, 29, 59]
	River water	DPASV/HMDE	[60]
Co	Sea water	ADPV/HMDE	[58, 60, 61]
	River water	ADPV/HMDE	[71, 62, 63]
Cr	Natural water	ADPV/HMDE	[64]
Cu	Sea water	DPASV/MFE	[57–58, 29, 65–66]
	River water	DPASV/HMDE	[71, 67]
Hg	Sea water	PSA/AuFE	[68]
Mo	Natural water	ADPV/SMDE	[69]
	Sea water	ADPV/HMDE	[70]
Ni	Sea water	ADPV/HMDE	[57, 58, 60]
	River water	ADPV/HMDE	[71, 72]
Pb	Sea water	DPASV/MFE, HMDE	[57, 58]
	River water	DPASV/HMDE	[71]
	Sea water	ADPV/HMDE	[72]
Pt	Sea water	ADPV/HMDE	[78]
Sn	Natural water	DPASV/HMDE	[79]
	Sea water	ADPV/HMDE	[80]
Ti	Sea water	ADPV/HMDE	[81, 82]
U	Sea water	ADPV/HMDE	[83]
	Polluted water	ADPV/HMDE	[84]
	Natural water	ADPV/HMDE	[85, 86]
	Tap water	FIA ADPV/MFE	[87]
Zn	Sea water	DPASV/HMDE	[36]
	River water	DPASV/HMDE	[71]
	Sea water	ADPV/HMDE	[29]

Other disadvantages of the two wet digestion techniques are the high consumption of chemicals, the possible loss or contamination of certain elements and poor reproducibility. Moreover, samples with fat content higher than 10% (dry weight) cannot be digested by these techniques.

For samples with high silica content (e.g. soil or sediment), simple extraction with dilute acid or complete digestion with the addition of hydrofluoric acid can be used to obtain solutions for direct electrochemical determination. Table 14.2 presents some examples for the electroanalytical determination of elements in inorganic samples after different wet digestion procedures. In industrial electrolytes it is possible in some cases to determine elements without sample decomposition.

14.6
Pressurized Digestion

Pressurized digestion is preferred to be the standard digestion method in all laboratories which determine elements in environmental and biological samples. Good reproducibility, low risk of contamination or loss of analyte, and well defined physico-chemical parameters are the advantages of these systems.

One of the most prominent decomposition methods during the past years has been the wet chemical decomposition with nitric acid in PTFE bombs. Bomb decomposition offers the great advantage that many substances which are inert towards nitric acid at lower temperatures may be attacked successfully at higher temperatures; an additional bonus is that the loss of volatile reaction products is avoided, unless of course there is an escape of gases when the bomb is opened after cooling. This method, however, is hampered by a major disadvantage: the decomposition temperature must not exceed 200°C because PTFE would lose the mechanical stability at high-temperatures. Frequently, the decomposition process at this temperature proceeds sluggishly and is incomplete.

Because the decomposition vessels are limited in volume, only small amounts of sample can be digested. Handling the vessels under pressure and the limited number of vessels which can be heated at the same time are further disadvantages of this digestion technique.

Solution obtained by this pressurized digestion cannot be used for subsequent voltammetric determination. However, in some matrices it is possible to determine lead by PSA. If the digested solution is treated with perchloric acid, it is also possible to determine Zn, Cd and Pd by anodic stripping voltammetry. By cathodic stripping voltammery, it is possible to determine only Ni. If soil or sediment samples are digested in a PTFE-bomb by a mixture of nitric and hydrofluoric acids, due to high concentration of elements in these matrices, it is possible to determine some elements by electrochemistry after evaporation to dryness and appropriate dilution of the digest.

In certain cases, PTFE adsorbs elements from the solution or leaks them into the solution; as a result losses or contamination (e.g. Cr from PTFE) can occur.

In the past few years, the microwave digestion procedure has proved to be a very useful tool (see also Chapter 13). This technique has the benefit of rapid sample preparation and reduced contamination risk. However, in some matrices, after digestion with nitric acid, a precipitate was observed, and it was found that o-, m-, and p-nitrobenzoic acids (NBS) and other organic species present in these digests irreversibly complex with copper (but not zinc), leading to low values for copper by electrochemistry [10]. New technological developments in microwave digestion systems (such as temperature or pressure control in the digestion vessels) have made it possible to optimize this digestion process. New types of PTFE introduced for microwave digestion vessels eliminate the problems of Cr contamination or element loss. The introduction of vessels made of quartz into the microwave technique make it possible to increase the digestion pressure to about 80 bar, and the digestion temperature can be correspendingly increased to about 260 °C. If only nitric acid is used for the digestion, it will not be possible to digest all biological samples in the quality needed for subsequent voltammetric determination. The addition of perchloric acid significantly increases the quality of digestion, but regular use of this acid in the laboratory is extremely hazardous (see also Chapter 10). After microwave digestion with a mixture of nitic and perchloric acids, it is possible in principle to determine nearly all electrochemically detectable elements in environmental samples.

The sample size which can be digested by microwave systems is limited by mechanical stability of the digestion vessels. Very rapid chemical reactions at the beginning of the microwave disgestion process induce high pressure peaks in the vessels. Since these peaks very much depend on the matrix involved, preliminary tests are necessary in order to optimize the microwave digestion programs for each matrix. In connection with a multi-element analytical technique such as ICP-MS [12], microwave digestion is an excellent method for decomposing materials with a high silica content.

Figure 14.3 shows the influence of the digestion temperature on the carbon content in the digestion solution if only nitric acid is used. Because the volume of the

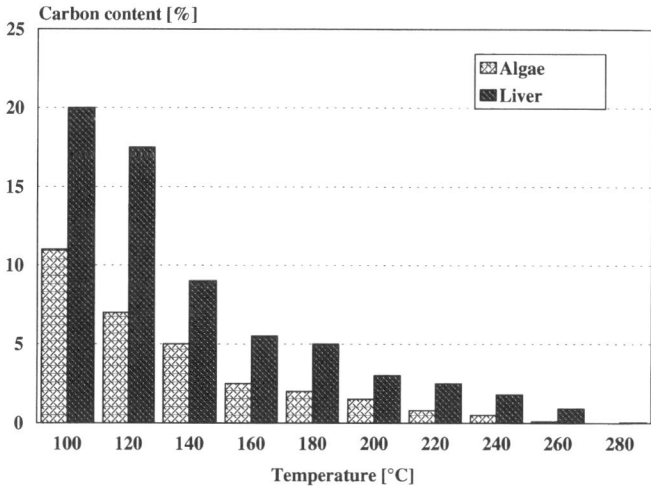

Fig. 14.3. Influence of the digestion temperature on the carbon content in the digestion solution if only nitric acid is used

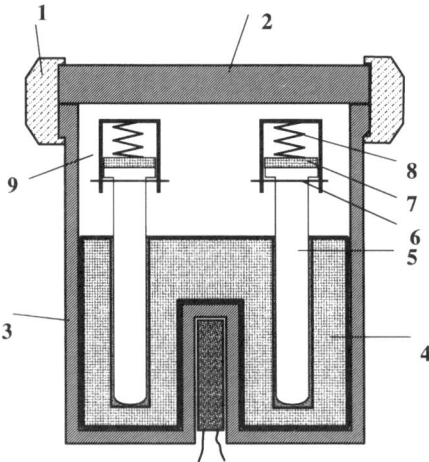

Fig. 14.4. High pressure ashing system: (1) shutter; (2) lid; (3) autoclave; (4) aluminium heating block; (5) sample vessels made from quartz; (6) tungsten clasp; (7) quartz lid; (8) tungsten spring; (9) cap

digestion vessel, the amount of acid used and the sample size are constant an increase in the digestion temperature causes the pressure in the digestion vessel to increase.

To avoid the problem of the loss of mechanical stability at high temperatures, vessels made of quartz are now being used in a new pressure decomposition system (high pressure ashing, HPA) [13]. These quartz vessels are covered by a quartz lid and then mounted in an autoclave (Fig. 14.4), equipped with a heated aluminium block and subjected to a pressure of 100 bar. A gas regulator unit permits easy filling and emptying of the autoclave, e.g. from a nitrogen tank.

Different size of vessels (2–70 ml) are available for optimally decomposing samples that vary in weight from a few mg to 1.2 g. Depending upon the size of the digestion vessel, up to seven units can be placed inside the autoclave.

A microprocessor temperature control allows a wide variety of temperature programs (Table 14.5) to be used. The adaptation of the temperature program to the specific decomposition task is very important. Many different organic materials (such as biological samples, polymers, coal, oil, fat and other organic industrial products) can be decomposed by this method.

Because organic substances can be completely decomposed, this digestion method is particularly suitable for electrochemical determination. It is possible to determine Zn, Cd, Pb, Cu, Ni and Co in different matrices using HMDE or MFE with different electrochemical detection methods. Low blanks for these elements by HPA digestion make it possible to determine these elements even at very low concentration levels, as observed in some environmental matrices (e.g. fish). If Tl is to be determined in environmental samples, due to its low concentration (< 50 µg/kg dry weight), 0.4 to 1 g of dry sample can be decomposed in a 70 ml digestion vessel. A special analytical procedure must be developed, according to the nature of matrix and the observed concentration range. During evaporation of nitric acid from digestion solution, loss of Tl is possible.

Selenium can also be determined by CSV on HMDE in samples from marine ecosystems and in food materials. After HPA digestion with nitric acid only, selenium

Table 14.3. Determination of elements by electroanalytical methods after wet digestion

Element	Matrix	Determination method	References
Ag	Waste water	DPASV/CPE	[8]
As	Plants	DPASV/HMDE	[20, 72]
Bi	Blood	PSA/AuE	[73]
	Plants	DPASV/HMDE	[20, 72]
Cd	Soil	DPASV/HMDE	[53, 73]
	Blood	PSA/AuE	[73]
	Wine	DPASV/HMDE	[74]
	Meat	DPASV/HMDE	[75, 76]
	Plants	DPASV/HMDE	[20, 72]
	Plants	PSA/MFE	[77]
Co	Blood	CSV/HMDE	[78]
	Hair	CSV/HMDE	[55, 79]
	Meat	CSV/HMDE	[75, 76]
Cu	Blood	PSA/AuE	[73]
	Meat	DPASV/HMDE	[75, 76]
Fe	High purity materials	CSV/HMDE	[80]
Ge	Ore samples	CSV/HMDE	[81]
In	High purity materials	CSV/HMDE	[80, 82, 83]
Mo	Blood	CSV/HMDE	[84]
Ni	Hair	CSV/HMDE	[78, 79]
	Blood	CSV/HMDE	[78]
	Meat	CSV/HMDE	[75, 76]
Pb	Hair	PSA/MFE	[77]
	Wine	DPASV/HMDE	[74]
	Meat	DPASV/HMDE	[75, 76]
	Plants	PSA/MFE	[77]
Sb	High purity materials	CSV/HMDE	[85]
Se	Blood	CSV/HMDE	[86]
	Hair	DPASV/HMDE, AuE	[20, 87]
Sn	Geological samples	CSV/HMDE	[88]
Th	Nuclear samples		[89, 90]
	Ore samples	CSV/HMDE	[81, 91]
Tl	Wine	DPASV/HMDE	[92]
Zn	Wine	DPASV/HMDE	[74]

is present in the digestion solution in the oxidation state of Se(IV). This oxidation state is particularly suitable for electrochemical determination. The digestion solution only needs to be diluted with an appropriate supporting electrolyte. Selenium can be also determined on a gold electrode using PSA or ASV as a detection method. If a mixture of nitric and perchloric acids are used for the digestion, the oxidation state of selenium changes to Se(VI) and a reduction step must be included in the analytical procedure before electroanalytical determination, can take place.

Arsenic determination by electroanalytical methods is possible but complicated. In nature, both inorganic and organic arsenic compounds exist. Decomposition of arsenic inorganic compounds is easy and can be done without any problems. In the case of some organic arsenic compounds (arsenobetains), which are very re-

Table 14.4. Determination of elements by electroanalytical methods after pressurized decomposition

Element	Matrix	Determination method	References
Al	Blood	CSV/HMDE	[94]
	Plants	CSV/HMDE	[95, 96, 97]
As	Meat	CSV/HMDE	[98]
	Sediment	DPASV/AuE	[99, 100]
Cd	Hair	CSV/HMDE	[101]
	Meat	DPASV/HMDE	[11]
	Plants	DPASV/HMDE	[75]
	Fish	PSA/MFE	[102]
	Bivalves	DPASV/HMDE	[102]
Co	Meat	CSV/HMDE	[62]
	Plants	CSV/HMDE	[75]
	Plants	PSA/MFE	[68]
	Bivalves	CSV/HMDE	[103]
Cr	Plants	CSV/HMDE	[104]
Cu	Meat	DPASV/HMDE	[11]
	Plants	DPASV/HMDE	[75]
	Wine	DPASV/HMDE	[26]
Hg	Sediment	DPASV/AuE	[99, 100]
Mo	Meat	CSV/HMDE	[105]
Ni	Blood	CSV/HMDE	[106, 72]
	Urine	CSV/HMDE	[106]
	Plants	CSV/HMDE	[75]
	Plants	PSA/MFE	[107]
	Bivalves	CSV/HMDE	[103]
	Wine	CSV/HMDE	[26]
Pb	Blood	DPASV/HMDE	[75, 103]
	Meat	DPASV/HMDE	[11]
	Plants	DPASV/HMDE	[75]
	Bivalves	PSA/MFE	[102]
	Fish	PSA/MFE	[102]
	Bivalves	DPASV/HMDE	[103]
Pt	Urine	CSV/HMDE	[108]
Se	Blood	PSA/CFE	[113]
	Meat	CSV/HMDE	[98, 109]
	Meat	PSA/AuE, MFE	[68, 110]
V	Urin	CSV/HMDE	[111]
	Sediment	CSV/HMDE	[112]
Zn	Plants	DPASV/HMDE	[20, 75]
	Wine	DPASV/HMDE	[26]
	Meat	DPASV/HMDE	[11]

sistant, it is necessary to increase the digestion temperature to 320 °C or to use a mixture of nitric and perchloric acids. This procedure must be used for fish, bivalves and other samples from marine ecosystems. The electrochemical procedure for arsenic determination is similar to that for selenium, but the sensitivity is less.

Chromium determination after HPA digestion by cathodic stripping voltammetry is very difficult due to the fact that after this digestion the oxidation state of chromium is only + 3. For electrochemical determination it is necessary to oxidise

Decomposition Methods for the Electrochemical Determination of Elements

Table 14.5. Some recommended temperature programs for the high pressure ashing system (HPA) prior to electrochemical determination

Sample	Weight	Acid mixture	Temperature program				Remarks
			Step	°C	min	°C	
Wheat Coffee Milk powder	0.5 g	5 ml HNO_3 (65%) 1 ml HCl (70%)	1 2	80 280	15 90	110 280	70 ml quartz vessels clear, deep blue solution
Plant leaves	0.2 g	2 ml HNO_3	1 2 3 4	25 150 150 300	15 30 15 60	150 150 300 300	30 ml quartz vessels, deep green solution, SiO_2
Pine needles Wood	0.4 g	4 ml HNO_3	1 2 3 4	25 100 150 320	15 15 15 90	100 150 320 320	70 ml quartz vessels, deep blue solution, SiO_2
Algae Mussels	0.2 g	2 ml HNO_3	1 2 3 4	25 150 150 320	15 30 15 90	150 150 320 320	30 ml quartz vessels, deep green solution, SiO_2
Fish	0.4 g	4 ml HNO_3	1 2 3 4	25 150 150 320	15 30 15 90	150 150 320 320	70 ml quartz vessels, deep blue solution
Blood Serum	0.5 ml	2 ml HNO_3 0.05 ml $HClO_4$	1 2 3 4	25 150 150 300	15 30 15 60	150 150 300 300	30 ml quartz vessels, nearly colourless solution
Urine	4 ml	4 ml HNO_3	1 2 3 4	25 150 150 300	15 15 15 30	150 150 300 300	70 ml quartz vessels, nearly colourless solution
Oil	0.4 ml	4 ml HNO_3 0.05 ml $HClO_4$	1 2 3 4	25 150 150 320	15 30 15 90	150 150 320 320	70 ml quartz vessels, nearly colourless solution

chromium to the +6 oxidation state. In the presence of traces of organic substances or nitride oxides, it is not possible to obtain a stable Cr^{+6} solution. Another difficulty is the adsorption of chromium and some other elements onto the quartz walls. If iron is to be determined, the addition of some hydrochloric acid to the digestion mixture is necessary. For chromium determination perchloric acid must be added and for antimony determination the addition of hydrochloric and perchloric acids are necesarry if quartz vessels are used. In the future new procedures for the determination of other elements in environmental samples after HPA digestion must be developed in order to fully utilize the potential of this digestion technique.

14.7
Conclusions

New equipment and better methods have increased the possibility of determining a number of elements in environmental samples by electrochemisty. Since digestion equipment with good reproducibility and a high quality of digestion is now commercially available, it is possible to adapt the electrochemistry for determining trace elements in environmental samples.

High-pressure digestion with its large digestion temperature range is the most universal digestion system at present. This system is not only useful for electrochemical determination but it can be also used for multi-element analytical techniques such as ICP-OES or ICP-MS. In combination with pressurized digestion, it is possible at the present time to determine by electrochemistry the following elements in environmental samples: Zn, Cd, Pb, Cu, Ni, Co, Tl, Sn, Se, As, U, Fe, and V. For other elements appropriate procedures will still have to be developed.

14.8
References

1. Wang J (1995) Stripping Analysis: Principles, Instrumentation and Applications. VCH Publishers, Deerfield Beach, Florida
2. Coppeland TR, Skogerboe RK (1974) Anal Chem 46:1257A
3. Jagner D (1983) Trends Anal Chem 2:53
4. Wang J (1994) Analytical Electrochemistry. Verlag Chemie Weinheim
5. van den Berg CMG (1991) Anal Chim Acta 250:265
6. Nürnberg HW (1985) Fresenius Z Anal Chem 320:741
7. DIN-Norm 38406, Part 16 (1990)
8. Kolb M, Rach P, Schäfer J, Wild A (1992) Fresenius J Anal Chem 342:341
9. Dorten W, Valenta P, Nürnberg HW (1984) Fresenius Z Anal Chem 317:264
10. Pratt KW, Kongston HM, MacCrchan WA, Koch WF (1988) Anal Chem 60:2024
11. Würfels M, Jackwerth E, Stoeppler M (1987) Fresenius Z Anal Chem 329:459
12. Quevauviller Ph, Imbert IL, Olle M (1993) Microchim Acta 111:1
13. Knapp G (1985) Intern J Environ Anal Chem 22:71
14. Wang J, Li R, Huiliang H (1989) Elektroanalysis 1:417
15. Van den Berg CMG, Murphy K, Riley JP (1986) Anal Chim Acta 188:177
16. Greulach U, Henze G (1995) Anal Chim Acta 306:217
17. Hua C, Jagner D, Renman L (1987) Anal Chim Acta 201:263
18. Komorsky-Lovric S (1988) Anal Chim Acta 204:161
19. Bauer KH, Neeb R (1988) Fresenius Z Anal Chem 330:11
20. Ostapczuk P (1992) Clin Chem 38:1995
21. Almestrand L, Jagner D, Renman L (1987) Anal Chim Acta 193:71
22. Eschnauer HR, Ostapczuk P (1992) Die Weinwissenschaft 47:206
23. Eskilsson H, Heraldsson C, Jagner D (1985) Anal Chim Acta 175:79
24. Newton MP, Van den Berg CMG (1987) Anal Chim Acta 199:59
25. Hoyer B, Florence TM (1987) Anal Chem 59:2839
26. Marin C, Ostapczuk P (1992) Fresenius Z Anal Chem 343:881
27. Yang Y, Liu G (1990) Xiamen Daxue Xuebao, Ziran Kexueban 29:318
28. Wang J, Mahmoud J (1987) Fresenius Z Anal Chem 327:789
29. Ornella A, Maurizio A, Giovanni S, Corrado S, Eduardo M (1995) Anal Chim Acta 305:200
30. van den Berg CMG, Nimmo M, Abollino O, Mentasti E (1991) Electroanalysis 3:477

31. Goa Z, Li P, Wang G, Zhao Z (1990) Anal Chim Acta 241:137
32. Hasebe K, Yamamoto Y, Ohzeki K, Kambara T (1986) Fresenius Z Anal Chem 323:464
33. Wang J, Mannino S (1989) Analyst 114:643
34. Wang J, Lu J (1995) Talanta 38:1481
35. Wang J, Lu J, Taha Z (1992) Analyst 117:35
36. Hao Z, Vire J-C, Patriarche GJ, Wollast R (1988) Anal Lett 21:1409
37. Moreno MA, Marin C, Vinagre F, Ostapczuk P (1995) Heavy Metals in the Environment, Hamburg, September 1995. Eds.: RD Wilken, U Förstner, A Knöchel, p 341–344
38. Wang J, Sucman E, Tian B (1994) Anal Chim Acta 286:189
39. Wang J, Varughese K (1987) Anal Chim Acta 199:185
40. Raber G, Kalcher K, Neuhold CG, Talaber C, Koelbl G (1995) Electroanalysis 7:138
41. El-Shahawi MS, Abu Zuhri AZ, Kamal MM (1994) Fresenius J Anal Chem 348:730
42. Breyer Ph, Gilbert BP (1987) Anal Chim Acta 201:33
43. Adeloju SB (1991) Anal Sci 7:1099
44. Wang J, Zadeii J (1987) Talanta 34:909
45. Cleven R, Fokkert L (1994) Anal Chim Acta 289:215
46. Vandenbalck JL, Patriarche GJ (1987) The Science of the Total Environment 60:97
47. Cai Q, Khoo SB (1995) Anal Chim Acta 282:329
48. Dragic VV, Gary W van L (1994) Fresenius J Anal Chem 350:352
49. Mlakar K, Branica M (1989) Anal Chim Acta 221:279
50. Jin W, Shi S, Wang J (1990) J Electroanal Chem Interfacial Electrochem 291:41
51. Mrzljak RI, Bond AM, Cardwell TJ, Knight RW, Newma OMG, Champion BR (1994) Analyst 119:1057
52. Wang J, Lu J (1992) Analyst 117:1913
53. Kurayasu H, Inokuma Y (1988) Bunseki Kagaku 37:623
54. Adeloju SB, Tran T (1986) Analyst 111:1355
55. Huiliang H, Jagner D, Renman L (1988) Anal Chim Acta 207:17
56. Cai Q, Khoo SB (1993) Anal Chim Acta 276:99
57. Mart L, Nürnberg HW, Rützel H (1985) The Science of the Total Environment 44:35
58. Mart L, Nürnberg HW (1986) Marine Chemistry 18:197
59. Si Q, Xu S, Cai J, Lu R, Zhu Y (1992) Fenxi Huaxue 20:272
60. Donat JR, Bruland KW (1988) Anal Chem 60:240
61. Rerez-Pena J, Hernandez-Erito JJ, Herrera-Melian JA, Collado-Sanchez C, van den Berg CMG (1994) Electroanalysis 6:1069
62. Bebeki VI, Voulgaropoulos AN (1992) Fresenius J Anal Chem 342:352
63. Farias PAM, Ohara AK, Takase I, Ferreira SLG, Gold JS (1993) Talanta 40:1167
64. Golimowski J, Valenta P, Nürnberg HW (1985) Fresenius Z Anal Chem 322:315
65. Quentel F, Elleouet C, Madec C (1994) Electroanalysis 6:683
66. Culjak I, Mlakar M, Branica M (1995) Electroanalysis 7:64
67. Farias PAM, Fereira SLG, Ohara AK, Bastos MB, Goulard MS (1992) Talanta 38:1487
68. Eduardo P, Ostapczuk P (1994) Anal Chim Acta 293:55
69. Pelzer J, Scholz F, Henrion G, Heininger P (1989) Fresenius Z Anal Chem 334:331
70. Farias PAM, Ohara AK, Nobrega AW, Gold JS (1994) Electroanalysis 6:333
71. Weidenauer M, Lieser KH (1985) Fresenius Z Anal Chem 320:550
72. Zhang Z, Cheng Z, Cheng S, Yang G (1991) Talanta 38:1487
73. Camman K, Anderson JT (1982) Fresenius Z Anal Chem 310:45
74. Huiliang H, Jagner D, Renman L (1987) Anal Chim Acta 202:117
75. Schramel P, Hasse S, Knapp G (1987) Fresenius Z Anal Chem 326:141
76. Rozali bin Othman M, Hill JO, Magee RJ (1987) Fresenius Z Anal Chem 326:350
77. Ostapczuk P, Valenta P, Stoeppler M, Nürnberg HW (1983) Chemical Toxicology and Clinical Chemistry of Metals. Ed.: SS Brown, J Savory, Academic Press, London, New York ISBN 0-12-137520-X, p 61
78. Van den Berg CMG, Jacinto GS (1988) Anal Chim Acta 211:129
79. Weber G (1985) Fresenius Z Anal Chem 322:311
80. Van den Berg CMG, Khan SH, Riley JP (1989) Anal Chim Acta 222:43

81. Zhao Z, Cai X, Li P, Yang H (1986) Talanta 33:623
82. Li H, van den Berg CMG (1989) Anal Chim Acta 221:269
83. Van den Berg CMG, Nimmo M (1987) Anal Chem 59:924
84. Blázquez LC, Garcia-Moncó RM, Cabanillas AG, Misiego AS (1989) Fresenius Z Anal Chem 334:166
85. Sandres S, Henze G (1994) Fresenius J Anal Chem 349:654
86. Huang B, Zhang H, Pu G, Yin F, Zheng S, Yang H (1985) Anal Lett 18:279
87. Wu T, Xiang W, Zhang F, Deng J (1988) Analyst 113:1431
88. Wang EK, Sun W (1985) Anal Chim Acta 172:365
89. Sander S, Wagner W, Henze G (1995) Anal Chim Acta 305:154
90. Buldini PL, Ferri D, Nobili D (1988) Analyst 113:1317
91. Zhang XL, Wang LZ, Ma CS, Hu YH (1994) Chin Chem Lett 5:881
92. Eschnauer H, Gemmer-Colos V, Neeb R (1984) Z Lebensm Unters Forsch 178:453
93. Naumann R, Schmidt W, Hohl G (1994) Fresenius Z Anal Chem 349:643
94. Setiadji R, Wang J, Santanarios G (1993) Talanta 40:845
95. Sun C, Gao Q, Xi J, Xu H (1995) Anal Chim Acta 309:89
96. Cai Q, Khoo SB (1993) Anal Chim Acta 276:99
97. Karpiuk M, Politowicz M, Stryjewska E, Rubel M (1995) Fresenius J Anal Chem 351:693
98. Farias PAM, Ohara AK, Takase I, Ferreira SLG, Gold JS (1993) Anal Chim Acta 271:209
99. Hatle M (1987) Talanta 34:1001
100. Hasebe K, Kakizaki T, Yoshida H (1985) Fresenius Z Anal Chem 322:486
101. Zhang Z (1991) Mikrochim Acta 1:89
102. Zhang Z, Chen SZ, Lin HM, Zhang H (1993) Anal Chim Acta 272:227
Tripathi RM, Raghunath R, Khundekar RN, Mambi KSV (1994) Bull Electrochem 10:318
103. Ostapczuk P, Froning M, Stoeppler M (1989) Fresenius Z Anal Chem 334:661
104. Li C, James BD, Magee RJ (1989) Mikrochim Acta II (4–6):149
105. Adeloju SBO, Pablo F (1995) Electroanalysis 7:476
106. Mhajan RK, Kaur N (1993) Bull Electrochem 9:38
107. Eduardo P, Ostapczuk P (1993) Fresenius J Anal Chem 346:952 and 957
108. Adeloju SB, Bond AM, Swee NT, Guizhen W (1990) Analyst 115:1569
109. Breyer Ph, Gilbert BP (1987) Anal Chim Acta 201:23
110. Eskilsson H, Haraldsson C (1987) Anal Chim Acta 198:231
111. Ostapczuk P (1996) Croatica Chimica Acta (in press)
112. Greulach U, Henze G (1995) Anal Chim Acta 306:217
113. Jindal VK, Khan MA, Bhatnagar RM, Varma S (1985) Anal Chem 57:380

Chapter 15

Decomposition [of Materials] for Trace Analysis in the Cement Industry

WOLFRAM RECHENBERG and GEORG BACHMANN

15.1 Introduction

The solid siliceous and organic materials used in the cement industry normally have to be decomposed before they can be chemically analysed. The decomposition process may vary considerably – from simple extraction with water to the more elaborate and time-consuming techniques of fusion. Samples are therefore rarely analyzed for trace elements without pre-treatment, by which the solid matter is brought into solution. The final choice of the method for preparing a suitable solution of the material for the chemical analysis is decided by the elements that are to be determined and by the matrix that is to be destroyed. A method of decomposition that provides a suitable solution for the determination of a single element may not always be satisfactory, if the solution is to be used for the decomposition of a number of elements. A knowledge of the elements and/or compounds that are likely to be present in the material under examination is of great value to the analyst, as it will considerably aid his selection of a suitable method of decomposition.

This paper presents a survey of the different decomposition methods used for the analysis of various materials, fuels and products of the cement industry [1]. The effectiveness of these methods, which are presented in the following, has also been tested with different materials of other industries and the methods have partly been adopted in several VDI-working groups. An important prerequisite for reliable chemical analysis is representative and careful sampling (Chap. 9). The sample has to be prepared for the analysis by drying, crushing, grinding and homogenizing the material before it is decomposed. Sometimes it may even be necessary to mix average samples that are mass- or time-dependent from various single samples so as to produce a representative sample for the decomposition procedures described below. The sample is decomposed and brought into a state (normally a solution [2]) which is suitable for determining its constituents. An overview will be given of possible decomposition methods, and these will then be illustrated by selected examples.

15.2 Decomposition Methods

The aim of the various decomposition methods is to bring the substances that are to be analysed into a liquid form. Normally, this would mean solubilization [2], but decomposition by fusion is also a possibility.

Table 15.1. Decomposition methods: an overview

Method	Procedure	Example
1. Leaching	Water	NO_2, CrO_4^{2-}
	Acid	Hg
	Base	Anions
	Oxidizing agent	S^{2-}
	Reducing agent	S^{2-}
2. Dissolution	Pressure bomb	As, Be, Cd, Cr, Pb
	HNO_3 + HF	Ni, Tl, V, Zn
3. Combustion	Tölg device	As, Be, Cd, Cr, Pb
	Calorimetric bombs	Ni, Tl, V, Zn
	Wickbold device	Hg
4. Fusion	$Li_2B_4O_7$	F^-
5. Evaporation	Volatilization	Ce, Pb, Tl
	in tubes	Hg

Table 15.1 gives an overview of the decomposition methods commonly used: the method, details of the procedure in each case and applications. Sometimes it may be sufficient simply to extract the sample using water or another extraction agent such as an acid or base; method is used for determining the water-soluble portion of the nitrite in cements dusts [3].

The trace element thallium can be extracted from soils, by refluxing with an acid, in particular nitric acid [4]. An example for leaching with bases is the soda extraction, a method often used prior to the qualitative detection of anions and one which is also recommended before making a quantitative determination of anions by ion chromatography [5]. In some cases it may be necessary to use an oxidizing or reducing agent in combination with acids or bases. Using bromine water as the oxidizing agent, it is possible to oxidize the entire sulfur content of iron pyrites, thereby converting it to iron sulphate. Similarly, using zinc and hydrochloric acid in combination with metalic chromium, it is possible to expel the sulphur in the form of hydrogen sulfide. Without the addition of metallic chromium, the decomposition would be incomplete [6]. Arsenic is a typical example of an element, which requires an oxidizing agent in the initial stage for complete decomposition to take place, followed by the addition of a reducing agent as a second step [7]. All the methods mentioned so far have one thing in common, namely that the matrix materials do not need to be decomposed completely. Often the matrix even remains largely undecomposed although the constituent to be determined is completely dissolved and thereby separated from the matrix.

If complete decomposition of the material is required, pressure may be used. Since 1955 numerous modified pressure systems have been described which use, an inner chamber made of a polytetrafluoroethylene (PTFE) vessel, enclosed in an outer container (bomb) of stainless steel or aluminium that is capable of withstanding high pressures. Such bombs are commercially available from Berghof [8], Bernas [9] and Parr [10]. They are the result of fundamental research done by Wahler [11] and Langmyhr and P.E. Paus [12], and have proven successful for the decomposition of silicate material [13]. At present the Autoclave-3 from Perkin-Elmer is favoured because of its easy handling. With this decomposition vessel, the origin and the behaviour of the trace elements arsenic, beryllium, cadmium, chromium, lead, nickel, thallium, vanadium and zinc during the burning process in cement manufacture have been determined [14–20].

These elements are also present in various fuels, which also had to be analyzed in order to calculate the mass balances. These samples were ashed using a Tölg-device [21] within different calorimetric bombs [22–24] or by means of the Wickbold combustion apparatus [25, 26]. If any ash remains after combustion, it has to be decomposed using a pressure bomb.

Various fusion agents for the decomposition of inorganic material are available [27]. Generally, decomposition by fusion is not used for trace element analysis since in most cases the liquid sample acquires contaminants during the fusion process. In addition, some elements may be lost as a result of the fusion [28, 29]. However, there are some exceptions. One of these is the determination of fluoride in which the test material is fused with lithium tetraborate [30]. In a gold/platinum vessel (5/95). This decomposition technique has also proved successful in the determination of major and minor elements by X-ray fluorescence analysis [31].

Various trace elements such as thallium may be separated from the matrix by evaporation using high temperatures and an inert gas propellant. This method of decomposition, known as volatilization analysis [32–37], can be used in a slightly modified form for the direct determination of mercury in the silicates and fuels of the clinker burning process [38].

15.3
Examples

Following this overview of possible decomposition methods, some special examples are given below. The elements chromium and mercury have been chosen to represent the leaching method. Solubilization will be discussed in more detail in connection with PTFE bombs. This will be followed by an overview of the combustion of fuels using several different techniques. The conditions for decomposition by fusion in the determination of fluoride will then be presented. Finally there will be some remarks on the volatilization method and a modification of this method for the determination of mercury.

15.3.1
Leaching with Water

Since deposited wastes may be leached by rain, a knowledge of the leaching procedure in the chemical analysis of wastes may play an important role in environmental protection.

The German Waste Disposal Act prescribes at least the partial reuse of waste material. For example, the residues from power stations and waste incinerators, such as slags or fly ashes, may be used together with cement as road construction material [38]. Before waste material is used in this way, two safety requirements

Table 15.2. Methods used for testing leachability

Designation	Preparation	Test
Method 1 DEV S4	Sample crushed	0/2 mm 100 g/l, 24 h
Method 2 DEV S4 modified	Sample crushed	5/10 mm 100 g/l, 24 h
Method 3 Flow through method	Sample unchanged	Proctor cylinder H/D = 120/96 mm

must be fulfilled: (a) that the incorporated elements willl not be leached out of the cement-bound material at all, (or at the worst only in small and harmless quantities), (b) that the material is resistant to wear and tear under outdoor conditions.

A standardized method for testing the leachability of metals from solid materials is a procedure called DEV S4, which is part of a German standard method (DIN 38 414, Part 4) for examining water, sewage and sludge [40]. Originally, this method was developed for the examination of unsolidified sludges and sediments.

According to the standard, the sludge should be tested unchanged in size. In practice 100 g of the crushed sample, which can be any material, is constantly shaken for 24 hours in 1 l of deionized water. The trace elements are determined in the filtrate. Though seldom mentioned, it is obvious that the leachability of an element greatly depends on the grain size of the material. In the following, a sample is considered to be prepared if it has a grain size distribution between 0 and 2 mm, according to DEV S4 (see Table 15.2). If the leaching procedure remains unchanged, but the grain size of the crushed material ranges from 5 to 10 mm, then the method is called "modified" according to DEV S4. The results of both these leaching methods are compared by means of a flow-through method [41].

With this method the leachability was tested on Proctor cylinders, which have a height of 120 mm and a diameter of 96 mm. Originally, the test material consisted of sand and cement, mixed to gether with water that contained several elements of environmental interest. In later experiments slags and fly ashes from a municipal waste incineration plant were used. An ordinary Portland cement is used to prepare the cement-bound, cylindrical bodies. The damp cylinder is placed into a cylindrical tube of acrylic glass and centered. The marginal gap between the inner wall of the tube and the outer surface of the cylinder is filled with a two-component resin in order to prevent any water flowing through the marginal zone during measurement. The test cylinder with its sample is then placed on the bottom of the testing device shown in Fig. 15.1 and closed hermetically with a cap. The device is filled with deionized water through the feed-and-discharge stopcock. In this way the air trapped inside the sample cylinder is pushed out. When the water reaches a level above the zero point of the meter, the stopcock is closed and nitrogen from a pressure bottle is introduced through the pressure valve. The pressure is raised to 1 bar and maintained. The measurement starts when the first water droplets can

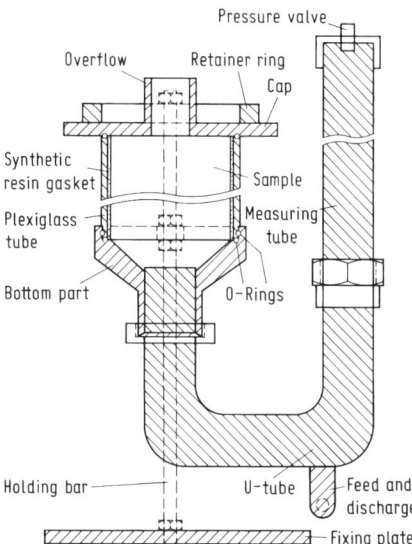

Fig. 15.1. Device for testing leachability

be seen on the top of the sample cylinder. Then the flow of the water through the sample cylinder is determined by metering the loss in height of water in the measuring tube over a certain time. From this measurement the permeability coefficient, k, of the sample can be calculated. Next, the water from the overflow is collected and analysed. The heavy metals are determined by atomic absorption spectrometry.

Figure 15.2 shows some results of the mortar samples that were investigated. Instead of fly ash and waste incineration ash, sand with an admixture of 6 to be 11% by weight of Portland cement was used. The mortars contained between 646 mg and 679 mg of chromium per kg of mortar in the form of water-soluble chromate. The curing time was 7 days (open circles) and 28 days (closed circles), respectively. The ordinate of the diagram shows the leaching rate of the heavy metals as a percentage. The abscissa gives the cement content of the mortars as a percentage by weight. The left part of the diagram shows the leaching rate of the crushed mortar sample with a maximum grain size of 2 mm, tested according to DEV S4. The middle part shows the results obtained with the modified DEV S4 procedure using coarser fragments of 5 to 10 mm diameter. The results of the flow-through method are presented in the right part of the figure.

Independent of the method of examination chosen, the leaching rates for chromium scarcely exceeded a value of 1% of the total heavy-metal content. Besides this, the leaching rate generally diminished with rising hydration, as may be seen from the differences between the open and closed circles. An interpretation of such behaviour would be that the consistency of, the test material changes, becoming denser in the course of time as more and more hydration products from which are able to bind a large portion of the heavy-metal content. The leaching rate also diminishes with rising cement content. Another important result needs to be emphasized: the magnitude of the leaching rate largely depends on the test method

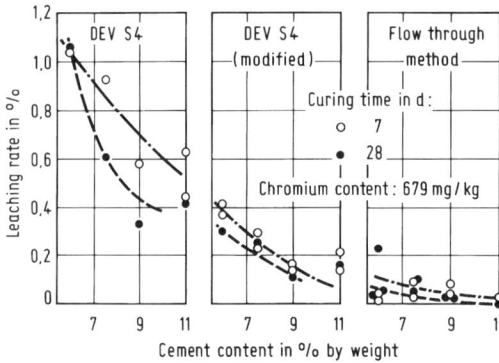

Fig. 15.2. Leaching rate of chromium depending on method, cement content and time of hydration

applied. It can be seen that the leaching rates for heavy metals, according to method DEV S4, were the highest because of the small grain size achieved by mechanical crushing of the material before it was subjected to the leaching procedure. By far the lowest rates were found by the flow-through method in which the original uncrushed solid material was used. In this last test, the heavy metals bound by the material will behave in the same way as they do under natural conditions. Only the results obtained for uncrushed materials are applicable to natural conditions, particularly when the leaching process is diffusion-controlled.

15.3.2
Leaching with Acids

Sinemus and Maier [42] cracked dissolved organic material bound to mercury in bottles made of polytetrafluoroethylene-perfluoropropylene co-polymer (FEP) that had been designed for the determination of mercury. Kaiser et al. [43] examied the loss of mercury when organic materials are oxidized using a mixture of hydrochloric and nitric acids. On the basis of these results, silicates of the cement-manufacturing process were tested to see whether they could be dissolved in FEP bottles for the determination of mercury. The bottles used had an outer diameter of 45 mm and a height of 110 mm. Nitric acid and hydrofluoric acid were added to the sample and the closed bottles were heated for up to five hours in a water bath at temperatures of 95 or 100 °C.

The results for three different standard reference materials are presented in Fig. 15.3. The resulting mercury content for the reference sample BCR 038 (coal fly ash) is plotted in g/t (ppm) in the upper part of the diagram, whereas the mercury contents of the light sandy soil (BCR 142) and the calcareous loam soil (BCR 141) versus the time of decomposition in hours ar plotted in the lower part of the diagram. For the latter two samples the mercury concentration is given in mg/t (ppb). The hatched areas in all three cases give the 95%-confidence intervals of the reference samples.

Although it is obvious that a decomposition time of one hour is insufficient for a total release of mercury from the reference material, it can be seen that the mer-

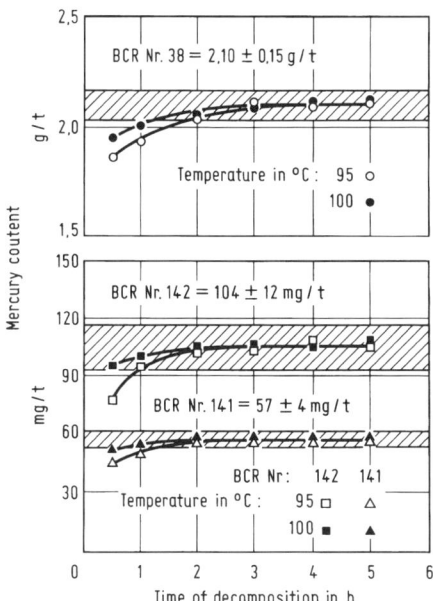

Fig. 15.3. Recovery of mercury from BCR standard reference materials as a function of decomposition time

cury content increases with the time of decomposition. After a heating period of at least three hours, the data points fall within the confidence intervals, indicating that the release of mercury is now complete [38]. The method has been used successfully for the determination of mercury in the silicates of the cement burning process [44]. In these applications it was found that the FEP becomes brittle in the course of time. Mercury may be adsorbed and desorbed by brittle FEP walls in the same way as for PTFE. In single cases a mercury concentration in brittle FEP of up to 1 g/t was found. To check whether an FEP bottle may still be used for analysis, a test run without a sample should be performed. If the mercury content of the solution is found to be lower than 0.01 µg Hg/l the bottle may be used again; otherwise the bottle is no longer suitable for mercury determinations [38].

15.3.3
Dissolution under Pressure

For the determination of trace elements in materials of the cement manufacturing process pressure decomposition methods are quite often used. Pressure bombs from Berghof, Bernas, Parr, and Perkin-Elmer are most commonly used. The containers and caps are made of polytetrafluoroethylene (PTFE). The casings are made of stainless steel or an aluminum alloy. The samples can be easily pressure-decomposed using a mixture of nitric and hydrofluoric acids to which – if necessary – other acids such as sulfuric acid or hydrochloric acid may be added. Generally, the decomposition time is 12 hours, and the temperature is maintained at 150 °C [13]. To avoid precipitation of poorly soluble calcium fluoride, a saturated solution of boric acid is added after the sample has been decomposed since this acid will react with the excess fluoride to form soluble tetrafluoroborate [9]. The result is a clear solution from which the elements arsenic, beryllium, lead, cadmium, chromium, nickel, thallium, vanadium and zinc can be determined by atomic absorption spectrometry [13, 18].

15.3.4
Combustion

Solid fuels such as lignite, coal, bleaching earth, rubber scrap and petroleum coke are incinerated in a Tölg device [21] (Fig. 15.4) made of silica glass. In the centre is the burning chamber with the sample holder and the inlet-system for oxygen or air. Sometimes it is necessary to use air instead of oxygen in order to slow down the combustion process. Above the burning chamber there is a cooling jacket and a cooling finger. Prior to the incineration, both are filled with liquid nitrogen which will condense the combustion vapours that form later. In addition there is also a water condenser above the cooling unit. Following the combustion, the apparatus is refluxed with nitric acid in a silica glass test tube. As a result of this most of the elements inside the device are collected in the test tube. However, the ashes of fuels with a high ash content remain in the sample holder. From there they are released mechanically, pulverized in an agate mortar and pressure-decomposed in a PTFE bomb (Sect. 15.3.3). After that the solutions from the two processes are mixed together.

During refluxing with nitric acid, it was noticed [13] that small quantities of nitric acid condensed near the inlet for air or oxygen. As a result of capillary action, this liquid crept towards the gas inlet and came into contact with a metallic tube, which is not shown in the figure. After the boiling procedure, a small amount of the nitric acid remained within the capillary tube. With the subsequent combustion, this

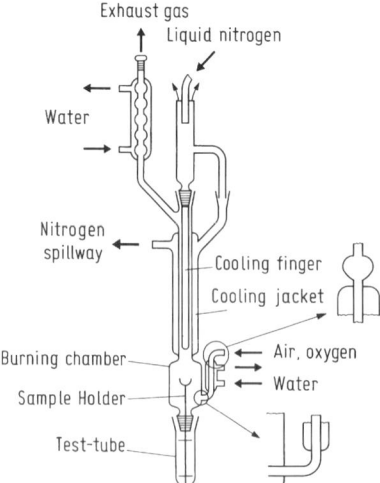

Fig. 15.4. Tölg-device for the combustion of fuels in a stream of air or oxygen

Table 15.3. Mercury content of fuels

Fuels	Repetitive determinations	Mean content in mg/kg	Standard deviation in mg/kg	Certified value in mg/kg
Coal 1	3	0.05	0.006	–
Coal 2	4	0.16	0.01	–
Coal 3	3	0.29	0.01	–
Lignite	5	0.07	0.008	–
Petroleum coke 1	4	0.04	0.01	–
Petroleum coke 2	5	0.08	0.004	–
Used tires	4	0.15	0.02	–
Coal NBS 1630	17	0.12	0.01	0.13 ± 0.01

nitric acid entered the solution of a later sample together with the air or oxygen, thus contaminating it and causing high values for chromium and nickel. To avoid this source of contamination, the metallic tube was replaced by a plastic hose. Furthermore the capillary tube (see the upper enlarged detail to the right of the figure) and the gas inlet (see the lower enlarged detail) were extended, thus preventing the nitric acid from entering the capillary [13].

Modified in this way, the apparatus gave good results for the combustion of lignite, hard coal or ballast coal. It can also be used to incinerate petroleum coke, rubber scrap and bleaching earth without any problems. The trace elements arsenic, beryllium, cadmium, chromium, lead, nickel, thallium, vanadium and zinc were fully recovered [13]; even mercury was quantitatively recovered, as can be seen from Table 15.3 in which the mercury concentrations (determined by atomic absorption spectrometry) of fuels decomposed in the Tölg device (Fig. 15.4) are given. The weight of each sample subjected to the combustion procedure was always about 0.1 g.

For these mercury determinations, 3 coal and 2 petroleum coke samples as well as 1 sample each of lignite and used tires were burned and analyzed. The determination was repeated three to five times. The mean values indicate a mercury content ranging from 0.04 (petroleum coke 1) to 0.29 g/t (coal 3). The corresponding standard deviations range from 0.004 (petroleum coke 2) to 0.02 g/t (used tires). The accuracy of the method was checked by analyzing the standard reference coal NBS 1630, which has a certified mercury content of 0.13 ± 0.01 g/t. The mean value of 17 determinations lay within the error limits of the certified value.

To avoid any loss of mercury while burning the fuels, the cooling system was repeatedly filled with liquid nitrogen [45]. In addition, after the combustion and the refluxing with nitric acid is finished, the apparatus has to be cleaned by flushing 5 to 10 ml of a 0.2% nitric acid solution through the gas outlet [45].

In some cases, for example gas coals or oil, some of the material was blown throughout the apparatus without actually burning, with the result that the decomposition of the sample was always incomplete. To get round this problem, this type of fuel was burned in calorimetric bombs [13].

Figure 15.5 fives a schematic impression of an oxygen combustion bomb. The bomb consists of a stainless steel container, closed at the top by a screw cap. The

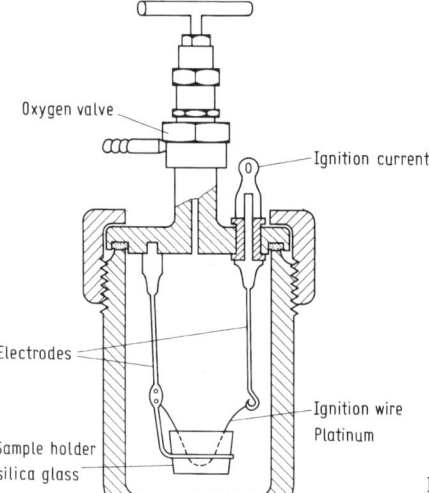

Fig. 15.5. Calorimetric bomb for the combustion of fuels

sample, wrapped in cigarette paper, is put into the silica glass sample holder and connected with the ignition wire. The metal of the ignition wire should not of course be (or contain) one of those elements that are being determined. Platinum, for example, is the most convenient metal for the determination of arsenic, beryllium, lead, cadmium, chromium, nickel, thallium, vanadium and zinc [13, 18]. The bomb is assembled, closed and filled with oxygen at a pressure of 40 bars. Afterwards the sample is ignited electrically, After a few seconds the combustion is over. The bomb is then cooled and opened. The inside of the bomb is washed out several times with water and all the washings are combined together. In the case of fuels with a high ash content, residues are left in the sample holder. These are collected and decomposed in PTFE bombs (see Sect. 15.3.3).

The bomb shown in Fig. 15.5 may be used without reservations for the determination of arsenic, beryllium, cadmium, lead, thallium and zinc. However, it is not recommended for the determination of chromium, nickel and vanadium because these elements will be released from the container itself, thus causing contamination and falsifying the results. These elements can be determined successfully using a bomb which contains an additional silica-glass beaker with a silica-glass cap, designed to prevent contamination of the solution by the bomb [18, 24]. The electrodes are also made of platinum so as to avoid contamination. So far no contamination by chromium, vanadium and nickel has been observed when the test solutions are collected in these silica-glass beakers.

Figure 15.6 shows a Wickbold combustion device [25, 26], which is made entirely of silica glass and consists of a vaporization chamber, an oxyhydrogen gas burner, a water-cooled burning chamber, a three-neck flask as a condensate collector and a water-cooled absorption tube.

20 ml of a dilute solution of potassium permanganate, acidified with sulfuric acid, are poured into the absorption tube. Then a sample of 100 to 500 mg in a pla-

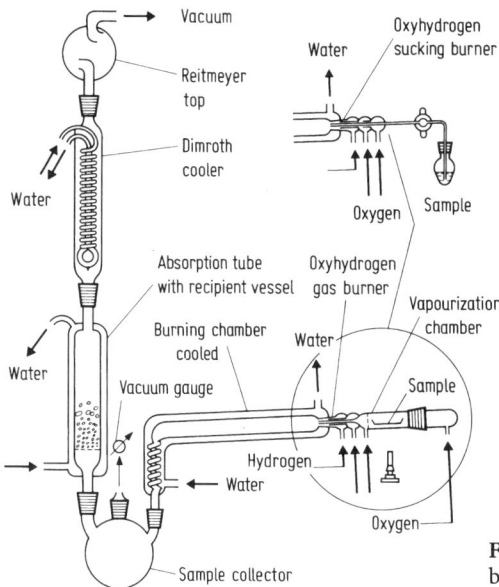

Fig. 15.6. Wickbold-device for the combustion of solid and liquid fuels

tinum boat is placed into the vaporizing chamber. The sample is heated from the outside by a gas burner until the fuel begins to smolder. It should not be coked or burned with an open flame. The process can be slowed down or accelerated by decreasing or increasing the oxygen flow. For this smoldering stage, a sharp oxygen-rich oxyhydrogen flame is recommended. After the organic matrix has been entirely destroyed, the remaining material in the platinum boat is strongly heated for another 10 minutes. The liquids in the absorption tube and in the condensate collector are combined. In this solution mercury can be determined by atomic absorption spectrometry using the cold-vapour technique. Further investigations have shown that mercury losses have to be taken into account if soot is deposited in the oxyhydrogen burner or in the burning chamber [45].

When liquid fuels, are to be burned, the vaporizing chamber is replaced by a suction burner as shown in the enlarged detail at the top right of Fig. 15.6. The fuel should be diluted with cyclohexane, sucked in and burned in the oxyhydrogen flame [26, 45].

15.3.5
Fusion

For the determination of fluoride, 0.1 to 1.0 g of the solid material is mixed with 4 g of lithium tetraborate. The mixture is fused in a gold/platinum crucible 5/95. After cooling, the melt solidifies to form a tablet which can be completely removed from the gold/platinum crucible by gentle tapping [30]. The tablet is transferred to the distillation flask of the device shown in Fig. 15.7 [46]. Perchloric acid and Aerosil,

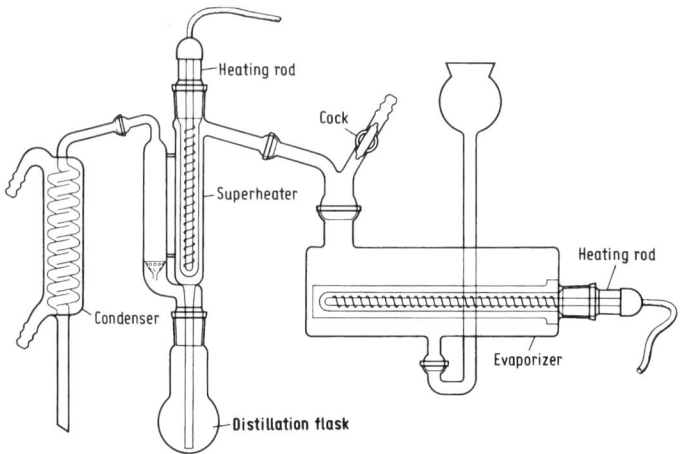

Fig. 15.7. Device for the separation of fluoride by superheated steam

a highly dispersed silicic acid, are added to the sample. The water vapour generated in the evaporator is passed through the superheater. The fluoride is subsequently completely separated off by the superheated water vapour and condensed. The fluoride in the distillate can be determined photometrically [30].

15.3.6
Evaporation

Some elements such as thallium, which form volatile compounds, can be evaporated from the matrix at elevated temperatures. The vapor is then transported by a propellant gas to an area of lower temperature where the volatile compounds are condensed [32–37]. A suitable device for this method is presented in Fig. 15.8. A sample of about 100 mg is heated in a platinum boat inside a silica-glass tube to a temperature of 1150 to 1200 °C in a nitrogen/hydrogen stream. The inner part of the tube is nozzle-shaped and directed centrally towards a cooling finger, in the inner part of which a reversed stream of cold water is constantly flowing. When these conditions are maintained, the thallium is evaporated completely out of the sample within 60 to 90 minutes and completely condensed in the trough of the cooling finger [37]. The lower part of the figure presents the temperature distribution within the tube, beginning from the middle of the oven and going to its outer rims. Open circles represent results obtained by Geilman [33], whereas closed circles represent our own values, which were obtained using the silica-glass tube, but without a propellant gas or a cooling finger. The difference between the two graphs probably results from the fact that in our experiments a more sophisticated and larger oven was used. When this is taken into account, the differences are negligible. Open triangles, on the other hand, represent the results we obtained when we used a propellant together with a cooling finger, as illustrated in the upper part of the figure. The temperatures were measured, avoiding any contact with the inner walls of the

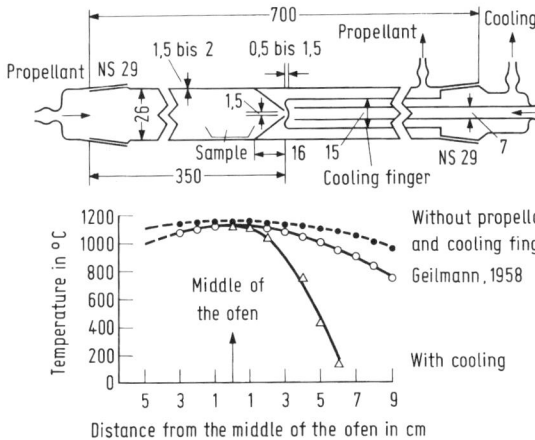

Fig. 15.8. *Top:* Silica glass tube for the separation of volatile elements from solid matrices. *Bottom:* Change of temperature as a function of distance from the center of the oven

tube. A curve drawn through these triangles indicates a gas temperature of about 800 °C near the trough of the cooling finger. The temperature on its surface, however, must obviously be lower because at 600 °C the thallium compounds would already be entirely in the gas phase [20]. At 400 °C only 10% will evaporate, whereas at a temperature of 330 °C the amount that evaporates is only 1%. Below a temperature of 100 °C, thallium can no longer be detected within the gas phase [47]. From these findings, it can be concluded that the temperature at the surface of the trough in the cooling finger is not likely to be significantly higher than 100 °C. This explains why the thallium condenses completely although the temperature of the oven is above 1150 °C.

The method is very time-consuming and is therefore not used for routine analyses. Nevertheless, it can be used for checking correctness of the results obtained for cadmium, lead and thallium after decomposition in a pressure bomb (Sect. 15.3.3).

Finally, a method quite often used in the cement industry for the direct determination of mercury needs be considered. In an oven with a nickel tube, mercury is evaporated and simultaneously determined by Zeeman atomic absorption spectrometry [48]. Approximately 0.03 to 200 mg of material are weighed and introduced into a pre-chamber, which is heated up to 1000 °C. The liberated mercury vapour is driven by a stream of oxygen into the nickel tube and measured. Before an unknown sample can be determined, a calibration line has to be established (Fig. 15.9). For this purpose, solutions of equal volume but with differing amounts of mercury are covered with calcium carbonate and atomized. Also various certified reference materials may be used, as can be seen in Fig. 15.9. From the individual weighings, the absolute amount of mercury can be calculated. In Fig. 15.9 the measurements are given in units of the signal area, wich is related to the absolute amount of mercury in nanograms. The open circles give the measurements obtained for aqueous solutions of mercury. The measurements for the various reference materials are shown by different symbols. The mercury content of these materials ranged from about 0.06 to 31 g/t. As may be seen from

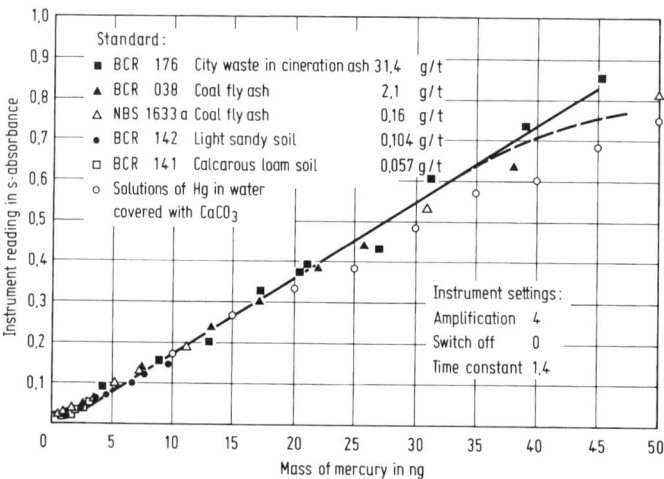

Fig. 15.9. Mercury calibration line from Hg solutions and standard reference materials for solid sample atomic absorption spectrometry

the figure, the relationship between the instrument readings and the amount of mercury is for the most part linear; with a downward bend towards the abscissa above about 0.4 reading units. A similar bend is also found for the range below about 0.1 units. Therefore, the readings for any samples should only be used in the range from 0.1 to 0.4 units.

The adjustments made to the instrument are also given in the figure. The given linearity is only valid for these adjustments. Since very small changes will shift the calibration line, it is necessary to repeat the calibration daily and check it frequently. If these precautions are taken, mercury in silicates and fuels can be accurately determined [38].

15.4
References

1. Working Group on analytical chemistry (1993) Determination of Trace Elements in Materials of the Cement Industry (German). Schriftenreihe der Zementindustrie, Heft 55. Beton-Verlag, Düsseldorf
2. Bock R (1972) Aufschlußmethoden der anorganischen und organischen Chemie. Verlag Chemie, Weinheim
3. Rechenberg W (1976) Zement-Kalk-Gips 29:254
4. Scholl W (1981) Landwirtschl. Forschung 34:275
5. Weiß J (1991) Ionenchromatographie VCH, Weinheim New York Basel Cambridge
6. Rechenberg W (1980) Zement-Kalk-Gips 33:126
7. Rechenberg W (1986) Zement-Kalk-Gips 39:155
8. Kotz L, Kaiser G, Tschöpel P, Tölg G (1972) Z Anal Chem 260:207
9. Bernas B (1968) Anal Chem 40:1682
10. Lechler PJ, Desilets MO, Cherne FJ (1988) Analyst 113:201
11. Wahler W (1964) N Jb Miner Abh 101:109

12. Langmyhr FJ, Paus PE
    Part 1 (1968) Anal Chim Acta 43:397
    Part 2 (1968) Anal Chim Acta 43:506
    Part 3 (1968) Anal Chim Acta 43:508
    Part 4 (1969) Anal Chim Acta 44:445
    Part 5 (1969) Anal Chim Acta 45:173
    Part 6 (1969) Anal Chim Acta 45:176
    Part 7 (1969) Anal Chim Acta 45:157
    Part 8 (1970) Anal Chim Acta 50:515
13. Rechenberg W (1986) In: Welz B (ed): Fortschritte in der atomspektrometrischen Spurenanalytik, Vol 2, Verlag Chemie, Weinheim Deerfield Beach FA-Basel, p 292
14. Sprung S, Rechenberg W (1978) Zement-Kalk-Gips 31:327
15. Sprung S, Rechenberg W (1983) Zement-Kalk-Gips 36:539
16. Rechenberg W (1983) Aus der Forschung – Für die Praxis, Research report No. 4. Verein Deutscher Zementwerke e. V. (ed), Düsseldorf
17. Sprung S, Kirchner S, Rechenberg W (1984) Zement-Kalk-Gips 37:513
18. Kirchner G, Rechenberg W (1986) In: Welz B (ed): Fortschritte in der atomspeltrometrischen Spurenanalytik, Vol 2, Verlag Chemie, Weinheim Deerfield Beach FA-Basel, p 299
19. Kirchner G (1985) Zement-Kalk-Gips 38:535
20. Kirchner G (1986) Das Verhalten des Thalliums beim Brennen von Zementklinker (Thesis). Schriftenreihe der Zementindustrie, Heft 47. Beton-Verlag, Düsseldorf
21. Morsches, B, Tölg G (1966) Z Anal Chem 219:61
22. Berthelot M (1892) Ann Chim Phys 26:555
23. Berthelot M (1892) Ann Chim Phys 26:559
24. Lindahl PC, Bishop AM (1982) Fuel 61:658
25. Wickbold R (1952) Angew Chem 64:134
26. Kulke M, Umland F (1977) Z Anal Chem 288:273
27. Dolezahl J, Povondra P, Sulcek Z (1968) Decomposition techniques in inorganic analysis. I liffe Books Ltd, London and American Elseviere Publishing Co., New York
28. Tölg G, Lorenz J (1968/1969) Methoden der mikrochemischen Elementbestimmung und ihre Grenzen. Fortschritte der chemischen Forschung, Vol 11. Springer-Verlag, Berlin Heidelberg New York
29. Tölg G (1975) Elemental analysis with minute samples. In: Svehla G (ed): Comprehensive analytical chemistry. Vol 3. Elsevier Scientific Publishing Co. Amsterdam Oxford New York
30. Rechenberg W (1972) Zement-Kalk-Gips 25:410
31. Richartz W (1971) Zement-Kalk-Gips 24:72
32. Geilmann W, Neeb R (1955) Angew Chem 67:26
33. Geilmann W (1958) Z Anal Chem 160:410
34. Geilmann W, Neeb K-H (1959) Z Anal Chem 165:251
35. Heinrichs H, Lange J (1973) Z Anal Chem 265:256
36. Heinrichs H (1975) Anal Chem 273:197
37. Walk H (1982) Die Gehalte der Schwermetalle Cd, Tl, Pb, Bi und weiterer Spurenelemente in natürlichen Böden und ihren Ausgangsgesteinen Südwestdeutschlands (Thesis), Karlsruhe
38. Bachmann G, Rechenberg W (1991) In: Welz B (ed) 6. Colloquium Atomspektrometrische Spurenanalytik, Bodenseewerk Perkin-Elmer, Überlingen, p 699
39. Sprung S, Rechenberg W (1988) Beton 38:193
40. DIN 38414, Teil 4 (10.84) Deutsche Einheitsverfahren zur Wasser-, Abwasser- und Schlammuntersuchung. Schlamm und Sedimente (Gruppe S). Bestimmung der Eluierbarkeit mit Wasser (S4). Beuth-Verlag, Berlin Köln
41. Rechenberg W, Sprung S (1987) In: Welz B (ed):4. Colloquium Atomspektrometrische Spurenanalytik, Bodenseewerk Perkin-Elmer, Überlingen, p 295
42. Sinemus HW, Maier D (1984) In: Welz B (ed) Fortschritte in der atomspektrometischen Spurenanalytik, Vol 1, Verlag Chemie, Weinheim Deerfield Beach FA-Basel, p 483
43. Kaiser G, Götz D, Tölg G, Knapp G, Maichin B, Spitzy H (1978) Z Anal Chem 291:278

44. Kirchartz B (1994) Reaktion und Abscheidung von Spurenelementen beim Brennen des Zementklinkers (Thesis). Schriftenreihe der Zementindustrie, Heft 56. Beton-Verlag, Düsseldorf
45. Bachmann G, Rechenberg W (1989) In: Welz B (ed) 5. Colloquium Atomspektrometrische Spurenanalytik, Bodenseewerk Perkin-Elmer, Überlingen, p 583
46. Seel F, Steigner E, Burger J (1964) Angew Chemie 76:532
47. Haegermann B (1982) Dampfförmige Schwermetallverbindungen im Zementofenabgas. MS Work, Clausthal
48. Bachmann G, Rechenberg W (1989) In: Welz B (ed): 5. Colloquium Atomspektrometrische Spurenanalytik, Bodenseewerk Perkin-Elmer, Überlingen, p 573

Subject Index

Aerosols/atmospheric 13
Ammonium ions
  precipitation 14, 22
  soils, soil solution 43
Analysis
  chemical clinical 8
  inorganic 8, 21, 33, 38, 40, 66, 67, 71, 96, 118
  organic 8, 78, 97
Analytical methods
  adsorption voltammetry 152, 168
  atomic emission 8, 21, 129
  cathodic stripping voltammetry 168
  chromatography 4
  cold vapour AAS 8, 193
  detection limits, overview 133
  electrochemical, overview 167–169
  graphite furnace AAS 8, 129
  hydride AAS 152
  ion chromatography 22
  matrix modification/AAS 8, 136
  plasma source mass spectrometry 21, 129
  potentiometric stripping 168
  properties, summary 133
  square wave voltammetry 151
  voltammetry 22, 129, 150
  X-ray fluorescence 185
Anions
  general 132
  in industrial materials 184, 193
  in rain 14, 22
  in soil 43
  in waste 69, 70
Aqueous samples
  centrifugation 10, 33, 34, 47, 52, 53
  conductivity 21
  filtration 4, 17, 22, 31–34
Arsenobetaine 129, 151

Bioindicators 3, 74, 88, 89
Biological samples (terrestrial)
  animal tissues 104, 105, 130
  earthworms 104
  mushrooms 130, 131
  pigeons 101
  poplar leaves 98, 99
  spruce needles 98, 99
  spruce shoots 95, 96

Clean room techniques 9, 27, 28, 105, 135
Cleaning procedures
  sample containers 80
  sampling bottles 28–30
  skin (human) 9
  vessels, sampling tools 2, 134
Contamination
  combustion 191
  decomposition 127–129, 133–135, 174
  dissection 92, 105
  fusion 185
  homogenization 92, 126, 127
  sampling 1, 2, 7, 26, 27, 30, 31, 34, 79, 92, 105
  slurry sampling 139
  storage 1, 92
  weighing 34

Decomposition general
  carbon content 129, 148, 152
  combustion/O_2 170–173, 190–193, 195
  comparisons 161, 162
  completeness 140–152, 162
  conditions 147–149
  evaporation 184, 194
  fusion 184, 185, 193, 194
  microwave assisted 155–166, 174, 175
  open wet 173, 174
  organic residues 162
  oxidative/ UV 169, 170
  pressurized 128, 129, 142–152, 174–179, 184, 185, 189, 190
  safety devices/features 146, 147, 157–159
  sample size/weight 129, 137, 138, 148, 149, 160, 175, 176, 193, 195
  temperatures 147, 151, 152
  time 147, 163–165
Decomposition/sea water 37, 38

Decomposition in analytical systems
  liquids   132, 136
  slurries   132, 136–139
  solids   132, 136–138, 195
Dissection/biological samples   2, 3, 84, 105
Dissolved organic carbon (DOC)   43
Dissolved organic matter (DOM)   37, 38, 169, 170

Elements
  Ag   71, 137, 171, 177
  Al   2, 4, 8, 14, 27, 31, 36, 38–40, 46, 109, 118, 124, 128, 135, 137, 138, 171, 173, 178
  As   2, 4, 8, 43, 69, 119, 123, 124, 128, 137, 138, 151, 152, 171, 172, 177, 178, 180, 184, 185, 190–192
  Au   27, 137
  B   128, 138
  Ba   95–97, 137
  Be   2, 4, 69, 119, 137, 138, 184, 185, 190–192
  Bi   137, 171, 177
  C   69, 148
  Ca   8, 14, 22, 43, 99, 109, 118, 128, 135–138
  Cd   2, 4, 8, 21, 22, 24, 29, 31, 38, 39, 43, 45, 46, 69–71, 96–99, 119, 122–125, 128, 130, 131, 135–138, 150–152, 169, 170–174, 176–178, 189, 184, 185, 190–192
  Co   2, 4, 10, 22, 69, 122, 128, 137, 138, 152, 170–173, 176–178, 180
  Cr   2, 4, 10, 43, 45, 69, 71, 123, 124, 128, 137, 138, 145, 171, 173, 174, 178, 179, 184, 185, 187, 188, 190–192
  Cu   2, 4, 5, 8, 22, 31, 38, 39, 43, 45, 46, 69–71, 96–98, 124, 128, 135, 137, 138, 150–152, 164, 170, 171, 173, 176–178, 180
  Fe   8, 27, 38, 39, 43, 69, 70, 128, 135–138, 145, 171, 177, 180
  Ga   137
  Ge   177
  Hg   2, 4, 8, 22, 44, 45, 69, 70, 122–125, 128, 130, 137, 151, 163, 173, 178, 184, 185, 188–196
  I   122, 177
  In   137
  K   8, 14, 22, 43, 99, 136–138
  La   137
  Li   137, 139
  Mg   8, 14, 22, 43, 96, 97, 99, 128, 135–137, 139
  Mn   2, 4, 8, 38, 39, 69, 128, 137, 139, 171
  Mo   71, 98, 122, 136, 137, 139, 171, 173, 177, 178
  Na   8, 14, 22, 128, 135–137, 139
  Ni   2, 4, 10, 22, 45, 71, 119, 124, 128, 135, 137, 139, 152, 164, 170–173, 176–178, 180, 184, 185, 190–192
  P   8, 69, 70
  Pb   2, 4, 5, 8, 21, 22, 24, 26, 27, 29–36, 38–41, 45, 69, 70, 96–99, 119, 122–125, 128, 135–137, 139, 150–152, 169–173, 177, 178, 180, 184, 185, 190–192
  Pd   136, 137, 139, 171, 177
  Pt   71, 171, 173, 178
  Rb   137
  Rh   71
  Ru   171
  S   8, 69, 70
  Sb   69, 71, 128, 151, 177
  Se   2, 4, 8, 10, 69, 122, 124, 137, 139, 151, 163, 171, 172, 176–178, 180
  Si   128, 135, 137, 139
  Sn   69, 71, 124, 171, 173, 177, 180
  Sr   8, 71
  Tc   69
  Th   69, 177
  Ti   46, 71, 128, 137, 139, 171, 173
  Tl   2, 4, 8, 69, 119, 125, 137, 139, 169–171, 173, 177, 180, 184, 185, 190–192, 194, 195
  U   171, 173, 180
  V   119, 139, 171, 178, 180, 184, 185, 190–192
  Zn   2, 4, 8, 10, 22, 27, 45, 46, 69–71, 96, 98, 99, 119, 124, 128, 137, 139, 150–152, 164, 171, 173, 174, 176–178, 180, 184, 185, 190–192
Errors
  analysis   5, 60
  frequency distribution   91
  metabolization   104
  pollutants in precipitation   24
  rain volume   17, 18
  sample conservation   92
  sample number/mass   92
  sample preparation   92
  sample storage   92
  sampling   5, 60–62, 91, 92
  surface adsorption   1, 7
  water filtration   32, 33

Freeze drying
  food   127
  soils   53, 54

Homogeneity/heterogeneity   3, 68, 137
Homogenization
  food   126, 127
  marine samples   82–85

Subject Index

technical material   108
Human material
  anamnestic questionnaire   8
  blood/blood plasma   9
  creatinine   9
  haemolysis   9, 10
  hair   10
  hematocrite   8
  milk   11
  proteine pattern   8
  real-time monitoring   7
  reference range   9
  urine   10

Industrial materials
  conveyor belts   13
  crude-gas dust   15
  fuel oil   115
  leachability test   186
  mass balance   109, 110, 119, 120
  mass flow   112, 118
  pneumatic conveyors   113, 114
  precipitated dust   116
  used tires   115
  volatile elements   117
  waste-gas-dust   116

Laws, regulations etc.
  food consumption recommendations/
    Germany   130, 131
  guideline values: Cd, Pb, Hg in food   124, 125
  maximum admissible metal levels (food)   124
  trace elements in and on food products   126
  German, ISO and EU regulations: waste   44, 57, 67–69, 185

Marine and limnic samples
  defaecation/mussels   81
  eelpout   81, 84
  fresh water   26, 27, 39
  herring gull eggs   81, 84
  macro algae   80, 84
  mussels   81, 84
  sea water   26, 27, 30–34, 37
  sediments   80, 83
  surface sea water   35, 36
  water column   35
  zebra mussels   100, 101

Organic compounds   8, 17, 96, 97

Precipitation/deposition
  deposition rates   88
  dry and wet deposition   99
  rain   13, 14–19
  snow   19–24
  zones   13
Purification, water, reagents   29

Quality assurance/quality control
  analysis   5
  decomposition   127–129
  proficiency testing   5
  sampling   5, 54, 55, 89–93

Radionuclides in food   130
Reference materials   40, 129, 130, 139, 150, 151, 188, 191, 195, 196
Results, examples
  Ba in spruce shoots   95
  concentration of five elements in solid wastes   69
  concentration of Pb in sewage sludges   64
  Hg content of fuels   191
  influence of digestion temperature on carbon content   175
  influence of the age (poplar leaves) on metal concentration   98
  long-term deposition of Cd and Pb   24
  long-term pH changes/precipitation   23
  species in gases from sewage sludges and waste deposits   72
  surface Pb concentration (Atlantic Ocean)   41
  variation of Pb in sewage sludge ash   71
  Zn in spruce shoots   96

Sampling (Sample collection)
  abiotic and biotic factors   100
  confidence interval   64, 65, 95–97
  documentation   51, 52, 78-80
  for speciation   4, 72
  number of samples   3, 62–65
  representativeness   27, 58–60, 63, 79, 80, 89, 90
  safety precautions   55
  sample size/mass/volume   3, 11, 85, 92, 95, 112, 114, 126
  sampling depth   49
  sampling duration   79
  sampling pattern   51
  sampling plan   2, 89, 93, 100
  sampling point/place   3, 10, 15, 49, 50, 80, 90, 94, 100, 111
  sampling report/protocol   51, 79
  sampling strategy   3, 14, 15, 27
  sampling time/period/age   3, 7, 10, 15, 79, 93, 98–100

Sampling
  sampling/tools, equipment   3, 9–11, 16–20, 28, 46, 65, 66, 80–85
  seasonal variations   80
  standardization   90
  statistical criteria   58–60
  weather conditions   79, 92, 99
Speciation   69, 72, 123, 132, 177, 179
Sample storage
  cryogenic   7, 53, 85
  rain   20, 21
  Soil solutions   54
Soil solutions   45, 47, 52, 53
Soils   48-52
Specimen banking   40, 63, 75–78, 94
Specimen bank/Germany, selected areas   77
Standard operation procedures (SOP)   7, 76–80, 89, 94, 96
Subboiling distillation   29, 128, 129, 134
Suspended particular matter   31, 34

Total organic carbon (TOC)   44

Trace element levels
  in body fluids   2
  in food   122, 123
  in plants and soils   4
  in sea and fresh water   2, 38, 39
  in soils, groundwater and soil solution   45
Trend monitoring   40, 41

Washing procedures
  fruit and vegetables   130
  human hair   10
  plant samples   102–104
Waste
  bulk sample   58, 59, 62, 63
  laboratory sample   58
  parent population   58, 64
  analysis   71, 72
  metal monitoring   69
  legal requirements   67–69
  theoretical considerations/sampling   58–65

Printing: Saladruck, Berlin
Binding: Buchbinderei Lüderitz & Bauer, Berlin